PAUL HEWITT
UNIV OF TOLEDO

# OXFORD MATHEMATICAL MONOGRAPHS

*Series Editors*

E. M. FRIEDLANDER   I. G. MACDONALD
L. NIRENBERG   R. PENROSE   J. T. STUART

# OXFORD MATHEMATICAL MONOGRAPHS

A. Belleni-Morante: *Applied semigroups and evolution equations*
I. G. Macdonald: *Symmetric functions and Hall polynomials*
J. W. P. Hirschfeld: *Projective geometries over finite fields*
N. Woodhouse: *Geometric quantization*
A. M. Arthurs: *Complementary variational principles* Second edition
P. L. Bhatnagar: *Nonlinear waves in one-dimensional dispersive systems*
N. Aronszajn, T. M. Creese, and L. J. Lipkin: *Polyharmonic functions*
J. A. Goldstein: *Semigroups of linear operators*
M. Rosenblum and J. Rovnyak: *Hardy classes and operator theory*
J. W. P. Hirschfeld: *Finite projective spaces of three dimensions*
K. Iwasawa: *Local class field theory*
A. Pressley and G. Segal: *Loop groups*
J. C. Lennox and S. E. Stonehewer: *Subnormal subgroups of groups*
D. E. Edmunds and W. D. Evans: *Spectral theory and differential operators*
Wang Jianhua: *The theory of games*
S. Omatu and J. H. Seinfeld: *Distributed parameter systems: theory and applications*
D. Holt and W. Plesken: *Perfect groups*
J. Hilgert, K. H. Hofmann, and J. D. Lawson: *Lie groups, convex cones, and semigroups*
S. Dineen: *The Schwarz lemma*
B. Dwork: *Generalized hypergeometric functions*
R. J. Baston and M. G. Eastwood: *The Penrose transform: its interaction with representation theory*
S. K. Donaldson and P. B. Kronheimer: *The geometry of four-manifolds*
T. Petrie and J. Randall: *Connections, definite forms, and four-manifolds*
R. Henstock: *The general theory of integration*
D. W. Robinson: *Elliptic operators and Lie groups*
A. G. Werschulz: *The computational complexity of differential and integral equations*
P. N. Hoffman and J. F. Humphreys: *Projective representations of the symmetric groups*
I. Györi and G. Ladas: *The oscillation theory of delay differential equations*
J. B. Griffiths: *Colliding plane waves in general relativity*
L. Evens: *The cohomology of groups*

# The Cohomology of Groups

LEONARD EVENS
*Department of Mathematics,
Northwestern University, Illinois*

Oxford   New York   Tokyo
CLARENDON PRESS
1991

Oxford University Press, Walton Street, Oxford OX2 6DP
Oxford New York Toronto
Delhi Bombay Calcutta Madras Karachi
Petaling Jaya Singapore Hong Kong Tokyo
Nairobi Dar es Salaam Cape Town
Melbourne Auckland
and associated companies in
Berlin Ibadan

Oxford is a trade mark of Oxford University Press

Published in the United States
by Oxford University Press, New York

© Leonard Evens 1991

All rights reserved. No part of this publication may be reproduced,
stored in a retrieval system, or transmitted, in any form or by any means,
electronic, mechanical, photocopying, recording, or otherwise, without
the prior permission of Oxford University Press

A catalogue record for this book is available from the British Library

Library of Congress Cataloging in Publication Data
Evens, Leonard.
The cohomology of groups/Leonard Evens.
p.  cm.
Includes bibliographical references and index.
1. Finite groups.  2. Homology theory.  I. Title.
QA171.E94  1991  512'.2—dc20  91-17140

ISBN 0-19-853580-5

Text keyed by the author using TeX
Printed and bound in Great Britain by
Bookcraft Ltd., Midsomer Norton, Avon

# Preface

Cohomology of groups is a specialized topic, but it has figured prominently in major developments in important areas of mathematics. Its roots lie in both algebra and geometry. The algebraic side of the theory began early in the 20th century in the work of Schur (1904, 1907, 1911). Schur studied what we now call $H^1(G, \mathbf{C}^*)$ and $H^2(G, \mathbf{C}^*)$ in the context of the theory of projective representations of groups. A careful reading of his papers reveals intimations of much of the modern theory. The Schur multiplier (Karpilovsky 1989), which today we denote $H_2(G, \mathbf{Z})$, is still an object of fundamental importance in group theory with many interesting problems yet to be resolved.

Schur's ideas were extended in the third and fourth decade of the 20th century in the theory of group extensions (Schreier 1926a, 1926b; Baer 1934) and in the theory of cross product algebras. The latter was important in the development of the part of number theory called class field theory. The foundations of class field theory was a subject of great interest during this period, and that theory was put in definitive form by Tate, making use of the cohomology of Galois groups, about 1950.

During the late 1930s and early 1940s, it came to be realized that these objects, arising in different areas, were cohomology groups of appropriate topological spaces. Mac Lane (1978) describes the history of this exciting period which ultimately led to the development of modern homological algebra. Many names are associated with important aspects of these developments. Mac Lane credits Hopf with beginning it all, but Hurevicz, Eckmann, Freudenthal, Eilenberg, Mac Lane, and others played important roles. Two basic papers (Eilenberg and Mac Lane 1947a, 1947b) were particularly influential in setting the algebraic theory in its modern form as an analogue of the topological theory.

A history of the subject since that time would merit a book of its own. It had ramifications in many different areas, and I will not pretend expertise in all of these. My interest in the subject started when I was a graduate student, working with John Tate, in the late 1950s. One of my main interests has been to relate the structure of the cohomology ring of a finite group to the structure of the group. Very little progress in this direction was made until Quillen (1971a, 1971b, 1971c) showed that the Krull dimension of the mod $p$ cohomology ring is the maximal rank of any elementary abelian $p$-subgroup. (This had been conjectured ten years earlier by Atiyah, Swan, and myself—with help from John Thompson.)

Quillen's proof introduced very powerful methods to the theory, and these were seized upon by Alperin (Alperin and Evens 1981) and others in the theory of 'varieties of modules' which exhibits deep and beautiful relations between the modular representation theory of the group and the structure of its mod $p$ cohomology ring. However, there is still much to be learned about how the structure of the group is reflected in its cohomology.

Group cohomology may be developed in a variety of ways, each with its strengths and weaknesses. The cohomology of a group $G$ is the cohomology of a topological space $B_G$ called the classifying space of that group, so the subject may be considered part of algebraic topology. Classifying spaces play an important role in algebraic topology. Many important results in group cohomology have particularly elegant proofs in this context. In this text, I have chosen to develop the subject from the perspective of *homological algebra*. My motivation was to reach certain important topics relatively quickly without assuming extensive background in algebraic topology or other subjects. However, I should warn the reader that such a limited perspective is not sufficient for a thorough understanding of the subject, particularly if one hopes eventually to make one's own contributions.

I assume familiarity with the basic concepts of homological algebra as developed in the texts on that subject by Cartan and Eilenberg, Hilton and Stammbach, or Mac Lane. In particular, I assume the reader is familiar with the definitions of Ext and Tor via resolutions, and may even have been introduced to some of the basic concepts in group cohomology. Of course, some familiarity with finite group theory is essential, although I tried to give references where possible for what I used. For Chapter 7, the reader will be better off if he or she has some experience with spectral sequences, but I have tried to summarize the basic facts I need to use. For the variety theory in Chapters 8 to 10, a basic grounding in commutative algebra will be helpful. I have found the text by Atiyah and MacDonald (1969) useful for this purpose, but there are several other excellent sources. My general philosophy about prerequisites has been to summarize as many of the basic facts as practical and otherwise to give explicit references where needed.

The general organization of the book is as follows. Chapters 1 to 4 develop the basics by building on concepts in homological algebra. The idea of using appropriate resolutions is exploited to develop the basic theory and to prepare the ground for later material about wreath products. Chapters 5 and 6 are concerned with the cohomology of wreath products and the theory of the norm map. The norm map is a kind of multiplicative transfer which has not been exploited fully in the literature, and perhaps this text will help to remedy that. Chapter 7 discusses the spectral sequence of a group extension. Chapters 8 to 10 develop variety theory and some of its applications. This material requires familiarity with basic commutative algebra, particularly the spectrum of a commutative ring. The exercises range from trivial verifications to somewhat more challenging problems. Some theorems needed later in the text are also included in the exercises, so the reader is advised to look at all of them and to do as many as feasible.

There are three kinds of spectra used in commutative algebra and algebraic geometry: the prime ideal spectrum, the maximal ideal spectrum when the base field is algebraically closed, and the space of algebra homomorphisms of the ring into an arbitrary field. These are roughly equivalent theories but differ in some details. The second alternative is popular since assuming the field is algebraically closed simplifies many issues. Quillen chose the third alternative, and that is in many ways the cleanest choice. I have chosen to use the prime ideal spectrum because I wanted to include the case of finite base field and also because it seemed to me that this choice was closest to the underlying commutative algebra. At certain points in the theory, the prime ideal spectrum does not suffice, so I resorted to the other theories through the use of so-called 'geometric points'. The reader should be aware of these distinctions when studying treatments by different authors since they may mean different things by the same terms, and sometimes the statements or proofs do not carry over from one approach to another without some elaboration.

Most of what appears in this book exists in the literature, but I have also included some previously unpublished results. For example, some of the results on the norm for abelian groups are new.

This book arose from a one quarter course I taught at Northwestern University, and previous versions have been used successfully as a qualifying examination topic for students in algebra and algebraic topology. If Northwestern were on the semester system, I might have written a somewhat more expansive book. Because of time constraints, I left out important results or just briefly hinted at them in exercises. Of course, the reader should not stop with this book, so I will recommend some other useful sources.

Brown's (1982) book provides a leisurely presentation of the elements of the subject and it develops much of the homological algebra needed. I heartily recommend it as a supplement to Chapters 1 to 4 of this book. Also, Brown gives an excellent treatment of many of the geometric aspects of the theory.

Two earlier books on the use of homological methods in group theory, one by Gruenberg (1970) and one by Stammbach (1973), contain interesting material essentially disjoint from anything discussed here.

I have made no attempt to discuss *Tate cohomology* which pieces together the cohomology groups in positive degrees with the homology groups in negative degrees (shifted down by 1). Tate cohomology plays a specially important role in class field theory and also in studying the actions of finite groups on topological spaces. See Brown (1982) or Chapter XII of Cartan and Eilenberg (1956) for treatments of this subject.

I have not even hinted at the use of group cohomology in algebraic number theory. Starting points for learning about that would be the texts by Serre (1965b, 1979) or Cassels and Frölich (1967).

The problem of understanding Chern classes of unitary representations of finite groups is still largely unsolved. The time may not yet be right for a definitive book on this subject, and I have not made any attempt to discuss it here. The text by Thomas (1986) provides a sampling of some of what is known on this subject.

Steenrod's definition of the reduced power operations (Steenrod 1962) played an important role in my development of the norm map, and the latter may be viewed as a generalization of the Steenrod operations. By appropriate use of the norm map, I was able to avoid the use of reduced powers in this text, but the reader should be aware that Steenrod's operations and the Steenrod algebra are of fundamental importance for many aspects of the subject. Much of this is connected to the cohomology of the symmetric group, another subject not discussed in detail in this book.

Chapters 8 to 10 of this book treat the theory of varieties of modules in some detail, but my emphasis tends to suppress the role of representation theory. Fortunately, Benson (1991) has written a comprehensive two volume work which covers all the representation theory and relevant group cohomology. He has also included several relevant topics from algebraic topology that would not otherwise be accessible to readers not conversant with that subject.

I owe a debt of gratitude to many people who contributed to my being able to write this book. Long ago, John Tate imbued me with certain points of view which I have undoubtedly incorporated in this text without due attribution. Jon Alperin got me involved with the mathematics leading to variety theory, which I suspect could have proceeded quite well without my participation, but I am glad that it did not. As is clear from the frequency with which their names appear in the text, David Benson and Jon Carlson were responsible for much of the mathematics I report on here, and they kindly shared many of their insights with me at early stages in their development. In particular, I found Benson's first book on the subject (Benson 1984) invaluable as a source in preparing this text. Mark Feshbach and Stefan Jackowski each contributed ideas which have simplified the exposition at crucial points. Finally, I should thank Leonard Scott for generous help and forbearance.

Many people helped by correcting errors or suggesting ways to improve the exposition. Particular thanks are due to Lisa Townsley Kulich, Anetta Bajer, and Joseph Riesen. For the final version of this book, I owe an enormous debt to Stephen Siegel. He went through the manuscript with a fine tooth comb correcting errors small and large, and he was responsible for significant changes in the text which clarified the logical organization of the material. Finally, I should note that this book was typeset using $\mathcal{A}\mathcal{M}\mathcal{S}$-

TeX 2.0. The entire mathematical community owes thanks to Donald E. Knuth and his followers who have provided powerful tools for producing finished works with minimal effort.

Evanston  L. E.
1991

# Contents

1 **Preliminaries**   1
    1.1 Definitions   1
    1.2 Note on sign conventions   4

2 **Explicit resolutions**   5
    2.1 Cyclic groups   5
    2.2 Free groups   6
    2.3 The bar resolution   7
    2.4 Minimal resolutions   11
    2.5 Building new resolutions from old resolutions   16

3 **Products in cohomology**   21
    3.1 Definitions   21
    3.2 Computations   25
    3.3 Connecting homomorphisms and Bocksteins   26
    3.4 The Universal Coefficient Theorem   29
    3.5 Cohomology rings of direct products and abelian groups   32

4 **Relations to cohomology of subgroups**   35
    4.1 Restriction and the Eckmann–Shapiro Lemma   35
    4.2 Transfer or corestriction   38

5 **Cohomology of wreath products**   45
    5.1 Tensor induced modules   45
    5.2 Wreath products and the monomial representation   46
    5.3 Cohomology of wreath products   49
    5.4 Odd degree and other variations on the theme   54

6 **The norm map**   57
    6.1 Definition of the norm map   57
    6.2 Proofs of the properties of the norm   59
    6.3 The norm map for elementary abelian $p$-groups   62
    6.4 Serre's theorem   64

## 7 Spectral sequences — 69
- 7.1 The spectral sequence of a double complex — 69
- 7.2 The LHS spectral sequence of a group extension — 72
- 7.3 Multiplicative structure in the spectral sequence — 80
- 7.4 Finiteness theorems — 87

## 8 Varieties and complexity — 93
- 8.1 The variety of a module — 93
- 8.2 Subgroups — 97
- 8.3 Relations with elementary abelian $p$-subgroups — 100
- 8.4 Complexity — 103

## 9 Stratification — 109
- 9.1 The Quillen stratification of $X_G$ — 109
- 9.2 Quillen's homeomorphism — 117
- 9.3 Avrunin–Scott stratification — 122
- 9.4 The rank variety — 124

## 10 Some related theorems — 131
- 10.1 The tensor product theorem and applications — 131
- 10.2 Varieties and corestriction — 138
- 10.3 Depth — 141

References — 147

Table of notation — 153

Index — 155

# 1
# Preliminaries

## 1.1 Definitions

Let $G$ be a group, and let $k$ be a commutative ring. We shall usually assume that $G$ is finite, and that $k$ is a Dedekind domain; the most important cases are $k = \mathbf{Z}$ or $k$ is a field. Let $kG$ denote the group ring of $G$ over $k$. All modules (except where noted) are to be *left* $kG$-modules. Since $G$ is a group, the anti-automorphism $kG \to kG$ defined by $g \mapsto g^{-1}$ provides an isomorphism between the category of left $kG$-modules and the category of right $kG$-modules, so the sidedness of modules is not usually an issue. Often, when the base ring $k$ is clear, we shall abbreviate $kG$ by $G$ in formulas.

Write $\mathrm{Hom}(M,N) = \mathrm{Hom}_k(M,N)$ and $M \otimes N = M \otimes_k N$. If $M$ and $N$ are $G$-modules, make $\mathrm{Hom}(M,N)$ and $M \otimes N$ into $kG$-modules by

$$(gf)(m) = gf(g^{-1}m) \quad \text{for } f \in \mathrm{Hom}(M,N), m \in M, g \in G$$
$$g(m \otimes n) = gm \otimes gn \quad \text{for } m \in M, n \in N, g \in G.$$

These are called the *diagonal* actions. More generally, if $M$ is a $G_1$-module and $N$ is a $G_2$-module, make $\mathrm{Hom}(M,N)$ and $M \otimes N$ into $k(G_1 \times G_2) \cong kG_1 \otimes kG_2$ modules by

$$((g_1 \times g_2)f)(m) = g_2 f(g_1^{-1}m) \quad \text{for } f \in \mathrm{Hom}(M,N), m \in M, g_i \in G$$
$$(g_1 \times g_2)(m \otimes n) = g_1 m \otimes g_2 n \quad \text{for } m \in M, n \in N \text{ and } g_i \in G.$$

Then the action of $G$ is obtained by composing the action of $G \times G$ with the diagonal homomorphism $\Delta : G \to G \times G$ ($g \mapsto g \times g$). This point of view will be useful when we discuss the product structures which play an important role in what follows.

If $M$ is a $G$-module, let $M^G$ denote the submodule of $M$ consisting of all elements *invariant* under the action of $G$. Then

$$\mathrm{Hom}(M,N)^G = \mathrm{Hom}_{kG}(M,N).$$

The dual functor of *coinvariants* is defined as follows. Let

$$I = \Big\{ \sum_{g \neq 1} a_g(g-1) \,\big|\, a_g \in k \Big\}$$

be the augmentation ideal of $kG$. (It is easy to see that $I$ is the kernel of the *augmentation* ring homomorphism $\epsilon : kG \to k$ defined by $\epsilon(g) = 1$ for $g \in G$.) Then $IM$ is the submodule of $M$ generated by all elements $[g, m] = gm - m$ for $g \in G$ and $m \in M$, and we shall sometimes denote it $[G, M]$. We define
$$M_G = M/IM = M/[G, M].$$
It is the largest factor module of $M$ on which $G$ acts trivially. ($G$ acts trivially when each element acts as the identity, i.e. when $kG$ acts through the augmentation $\epsilon : kG \to k$.)

By analogy, set
$$M \otimes_{kG} N = (M \otimes N)_G.$$
Observe that this is consistent with the usual definition of the tensor product of a right $kG$-module $M$ with a left $kG$-module $N$ since in effect it forces the relations
$$m \otimes n = g(m \otimes n) = gm \otimes gn = mg^{-1} \otimes gn.$$

By definition, $\mathrm{Ext}^*_{kG}(M, N)$ is the sum of the right derived functors of the bifunctor $\mathrm{Hom}_{kG}(M, N)$, and $\mathrm{Tor}^{kG}_*(M, N)$ is the sum of the left derived functors of the bifunctor $M \otimes_{kG} N$. In particular, define
$$H_n(G, M) = \mathrm{Tor}^G_n(k, M)$$
$$H^n(G, M) = \mathrm{Ext}^n_G(k, M).$$

Since tensoring is right exact, the exact sequence $0 \to I \to kG \to k \to 0$ shows that $M_G \cong k \otimes_{kG} M$, and similarly $M^G \cong \mathrm{Hom}_{kG}(k, M)$. Thus, $H_n(G, -)$ is the $n$th left derived functor of the right exact functor $M_G$, and $H^n(G, -)$ is the $n$th right derived functor of the left exact functor $M^G$.

In principle, these definitions seem to depend on the base ring $k$, but as we shall see later the ring homomorphism $\mathbf{Z} \to k$ defined by $1 \mapsto 1$ induces isomorphisms
$$\mathrm{Tor}^{\mathbf{Z}G}_n(\mathbf{Z}, M) \cong \mathrm{Tor}^{kG}_n(k, M) \tag{1.1}$$
$$\mathrm{Ext}^n_{kG}(k, M) \cong \mathrm{Ext}^n_{\mathbf{Z}G}(\mathbf{Z}, M) \tag{1.2}$$
where of course $M$ may be viewed either as a $k$-module or through $\mathbf{Z} \to k$ as a $\mathbf{Z}$-module.

Recall how $\mathrm{Ext}_{kG}(M, N)$ is defined. Choose a projective resolution
$$\to P_n \to P_{n-1} \to \cdots \to P_1 \to P_0 \to 0 \to 0 \to \cdots$$
of $M$ with augmentation $\epsilon : P_0 \to M$ making
$$\to P_n \to P_{n-1} \to \cdots \to P_1 \to P_0 \to M \to 0$$

an exact sequence. We abbreviate this $\epsilon : P \to M$. Similarly, we let $\eta : N \to Q$ denote an injective resolution of $N$. (By the usual convention, the homogeneous components of $Q$ will be denoted $Q^n$.) Then

$$\operatorname{Ext}^n_{kG}(M,N) \cong H^n(\operatorname{Hom}_{kG}(P,N))$$
$$\cong H^n(\operatorname{Hom}_{kG}(P,Q)) \cong H^n(\operatorname{Hom}_{kG}(M,Q)).$$

In the middle term, we have the the *double complex* $\operatorname{Hom}_{kG}(P,Q)$ to which there is associated a single complex, and the $H^n$ refers to the cohomology of that single complex. Similar remarks apply to Tor but in that case one uses *projective* resolutions for either or both of the arguments.

It is clear now how to demonstrate formulas (1.1) and (1.2). Suppose for example that $X \to \mathbf{Z}$ is a projective $\mathbf{Z}G$-resolution of $\mathbf{Z}$. Since a $\mathbf{Z}G$-projective module is $\mathbf{Z}$-free, the resolution may be formed by splicing together short exact sequences which split. Tensoring with $k$ preserves these splittings, so $k \otimes_\mathbf{Z} X \to k$ is an exact complex over $k$. Since $k \otimes \mathbf{Z}G \cong kG$, it follows that $k \otimes_\mathbf{Z} X \to k$ is a $kG$-projective resolution. The usual morphism $X \to k \otimes X$ defined by $x \mapsto 1 \otimes x$ is $\mathbf{Z} \to k$ semilinear, and it induces a map of complexes

$$\operatorname{Hom}_{kG}(k \otimes X, M) \to \operatorname{Hom}_{\mathbf{Z}G}(X, M)$$

which is easily seen to be an isomorphism. (What is its inverse?) It follows that the functorially induced morphism in (1.2) is an isomorphism. A similar argument works for Tor.

If one is to make use of these definitions, then one should know something about projectives over $kG$. For $k$ a field, $kG$ is an artinian ring (even a symmetric $k$-algebra), and quite a lot is known about such projectives. For example, the indecomposable projectives are all direct summands of $kG$. (See Benson (1984, Chapter 1).) For $k = \mathbf{Z}$ or more generally a Dedekind domain, the situation is much more complicated. (See Swan (1960a).)

By definition, $H_*(G, A)$ and $H^*(G, A)$ are functors. $H_*(G, A)$ is covariant in both variables while $H^*(G, A)$ is contravariant in $G$ and covariant in $A$. For cohomology, we may combine the variance as follows. Let $\phi : G' \to G$ be a group homomorphism. If $A$ is a $G$-module and $A'$ is a $G'$-module, then $A$ may be viewed as a $G'$-module through $\phi$. Suppose $f : A \to A'$ is a $G'$-module homomorphism, i.e. $f(\phi(x')a) = x'f(a)$ for $x' \in G'$ and $a \in A$. Then we may compose the induced homomorphisms $\phi^* : H^*(G, A) \to H^*(G', A)$ and $f^* : H^*(G', A) \to H^*(G', A')$ to obtain

$$(\phi, f)^* = f^*\phi^* : H^*(G, A) \to H^*(G', A').$$

There are two specially interesting cases. First, suppose $H$ is a subgroup of $G$ (usually denoted $H \leq G$), $A$ is a $G$-module and $A' = A$ with

$H$ acting as a subgroup. Let $\iota : H \to G$ denote the inclusion map, and let $f = \mathrm{Id} : A \to A$. Then $(\iota, \mathrm{Id})^* : H^*(G, A) \to H^*(H, A)$ is called *restriction*, and denoted $\mathrm{res}_{G \to H}$. Second, let $N$ be a normal subgroup of $G$ (usually denoted $N \triangleleft G$), and let $A$ be a $G$-module. Then $A^N$ is a $G$-submodule of $A$ and since $N$ acts trivially on it, $A^N$ becomes a $G/N$-module. Let $\pi : G \to G/N$ denote the map of elements to cosets, and let $f : A^N \to A$ denote the inclusion. Then $(\pi, f)^* : H^*(G/N, A^N) \to H^*(G, A)$ is called *inflation* and is denoted $\mathrm{inf}_{G/N \to G}$. There are no commonly used terms for the corresponding maps in homology.

**Exercise 1.1.1.**
(a) Review homological algebra by describing $(\phi, f)^*$ at the level of resolutions. Thus, let $X' \to k$ and $X \to k$ be $kG'$ and $kG$ projective resolutions respectively, and let $\Phi : X' \to X$ be a map of resolutions such that $\Phi(g'x') = \phi(g')\Phi(x')$. Show that $(\phi, f)^*$ arises from $\Psi : \mathrm{Hom}_{kG}(X, A) \to \mathrm{Hom}_{kG'}(X', A')$ where $\Psi(\alpha) = f \circ \alpha \circ \Phi$.
(b) Show how to define a map $H_*(G, A) \to H_*(G/N, A_N)$ which is an analogue of inflation.

## 1.2 Note on sign conventions

When dealing with complexes various signs are introduced, but unfortunately different authors are not consistent in their usage. We shall follow Cartan and Eilenberg (1956, Section IV.5). We describe their sign convention for the functor Hom. A similar description applies to the functor $\otimes$.

Let $X = \{X_p, d_p : X_p \to X_{p-1}\}$ denote a (chain) complex and $Y = \{Y^q, d^q : Y^q \to Y^{q+1}\}$ a (cochain) complex. Then the two differentials in the double complex $\mathrm{Hom}(X, Y)$ are defined as follows:

$$d'^{p,q} = \mathrm{Hom}(d_{p+1}, \mathrm{Id}) : \mathrm{Hom}(X_p, Y^q) \to \mathrm{Hom}(X_{p+1}, Y^q),$$
$$d''^{p,q} = (-1)^p \mathrm{Hom}(\mathrm{Id}, d^q) : \mathrm{Hom}(X_p, Y^q) \to \mathrm{Hom}(X_p, Y^{q+1}).$$

In particular, note that there is no sign at all if $X$ or $Y$ is just a module, i.e. of degree 0. The general principle is that whenever a morphism of degree $r$ is moved past an argument of degree $s$, you should multiply by $(-1)^{rs}$.

Mac Lane (1963) and others use a different convention which introduces the sign $(-1)^{n+1}$ for $\mathrm{Hom}(X_n, M) \to \mathrm{Hom}(X_{n+1}, M)$.

# 2
# Explicit resolutions

It is sometimes possible to calculate cohomology efficiently by a clever choice of resolution. This observation was originally motivated by the fact that calculating the cohomology of a space is often facilitated by choosing an appropriate cellular decomposition. (See Brown (1982, Chapter 1).) However, it often possible to approach the issue purely algebraically.

## 2.1 Cyclic groups

Let $G = <g>$ be cyclic with generator $g$, and suppose $|G| = q$. Construct a free resolution of $k$ as follows. For each $n$ let $X_n = kGx_n$, and let

$$\epsilon(x_0) = 1$$
$$d_n(x_n) = (g-1)x_{n-1} \quad \text{for } n \text{ odd}$$
$$= Tx_{n-1} \quad \text{for } n \text{ even, } n > 0,$$

where $T = 1 + g + g^2 + \cdots + g^{q-1}$. It is clear that $dd = 0$, and it is not hard to see directly that the complex is acyclic (i.e. the augmented complex is exact). However, another way to see this is to make use of the concept of a *contracting homotopy* (Hilton and Stammbach 1971, Section IV.3). In this case, the $k$-module maps $\eta : k \to X_0$ defined by

$$\eta(1) = x_0$$

and $s_n : X_n \to X_{n+1}$ defined by

$$s_n(g^k x_n) = \begin{cases} (1 + g + g^2 + \cdots + g^{k-1})x_{n+1} & 0 < k < q \\ 0 & k = 0 \end{cases} \quad \text{for } n \text{ even}$$

$$s_n(g^k x_n) = \begin{cases} 0 & 0 \le k < q-1 \\ x_{n+1} & k = q-1 \end{cases} \quad \text{for } n \text{ odd}$$

provide such a contracting homotopy. This is summarized in the diagram

$$\begin{array}{ccccccccccc} \to & X_{n+1} & \to & X_n & \to & X_{n-1} & \to \cdots \to & X_1 & \to & X_0 & \to k \to 0 \\ & \uparrow & \nwarrow \uparrow & & \nwarrow \uparrow & & \nwarrow & \nwarrow \uparrow & \nwarrow \uparrow & \nwarrow \uparrow \\ \to & X_{n+1} & \to & X_n & \to & X_{n-1} & \to \cdots \to & X_1 & \to & X_0 & \to k \to 0 \end{array}$$

5

in which

$$\epsilon\eta = \mathrm{Id}$$
$$\eta\epsilon + d_1 s_0 = \mathrm{Id}$$
$$s_{n-1}d_n + d_{n+1}s_n = \mathrm{Id} \quad \text{for } n > 0.$$

It is easy to check the acyclicity of the complex from these relations. In making explicit calculations a contracting homotopy is often useful.

Since $\mathrm{Hom}_{kG}(kG, M) \cong M$, it follows that the complex $\mathrm{Hom}_{kG}(X, M)$ may be represented by

$$\cdots \to 0 \to M \xrightarrow{g-1} M \xrightarrow{T} M \xrightarrow{g-1} M \xrightarrow{T} M \xrightarrow{g-1} \cdots$$

so that

$$H^0(G, M) = M^G$$
$$H^n(G, M) = M^G/TM \quad \text{for } n \text{ even and } n > 0$$
$$H^n(G, M) = {}_TM/IM \quad \text{for } n \text{ odd,}$$

where ${}_TM = \{m \in M \mid Tm = 0\}$. These isomorphisms do not indicate how the cohomology groups behave as functors of the argument $G$, and that is a matter of some importance when making calculations.

**Exercise 2.1.1.** Show that if $G$ is a finite cyclic group and $M$ is a $G$-module, then $|G|H^n(G, M) = 0$ for $n > 0$. Hint: For $n$ odd, consider $|G|x - Tx$ for $x \in M$.

## 2.2 Free groups

Let $F$ be a free group on a generating set $S$. Then the augmentation ideal $I_F$ in $kF$ is free with basis $\{s - 1 \mid s \in S\}$. To see this one argues as follows. For any group $G$ and $G$-module $M$, a *derivation* $f : G \to M$ is a function satisfying

$$f(xy) = f(x) + xf(y).$$

(Such a function is also called a *crossed homomorphism*.) In particular, we have the *universal* derivation $D : G \to I_G$ defined by $D(g) = g - 1$. It is easy to see that any derivation $d : G \to M$ factors uniquely through $D$:

$$\begin{array}{ccc} G & \xrightarrow{D} & I_G \\ {\scriptstyle d} \searrow & \swarrow {\scriptstyle h_d \in \mathrm{Hom}_{kG}(I_G, M)} \\ & M & \end{array}$$

It follows that $d \mapsto h_d$ defines an isomorphism $\mathrm{Der}(G, M) \cong \mathrm{Hom}_{kG}(I_G, M)$ of functors on $M$. In the case when $G = F$ is free on $S$, any function $f : S \to M$ may be extended (necessarily uniquely) to a derivation $d_f : F \to M$. ($d_f$ may be defined inductively using the fact that any element of $F$ may be written uniquely as a *reduced word* $y_1 y_2 \ldots y_t$ where $y_i$ or $y_i^{-1} \in S$.)

It follows from the above discussion that $I_F$ is free over $kF$ with basis $\{s - 1 \mid s \in S\}$, so

$$0 \to \bigoplus_{s \in S} kF(s-1) = I_F \to kF \to k \to 0$$

is a free resolution of $k$. Hence,

$$H_n(F, M) \cong 0 \quad \text{for } n > 1$$
$$H_1(F, M) \cong \{\sum_{s \in S} m_s \in \bigoplus_{s \in S} M \mid \sum (s-1)m_s = 0\}$$
$$H_0(F, M) \cong M/[F, M].$$

See Hilton and Stammbach (1971, Section VI.5) or Brown (1982, Section IV.2) for more discussion of these points.

One very important case is $F \cong \mathbf{Z}$ free on one generator $s$. In that case

$$H_1(F, M) = M^F = \{m \in M \mid sm = m\}.$$

If $G = F/R$ is presented as a free group modulo a normal subgroup of relations where $F$ is generated by $S$ and $R$ is generated *as a normal subgroup* by $T$, then one can start a free resolution of $k$ by a sequence

$$\bigoplus_{t \in T} kG \to \bigoplus_{s \in S} kG \to kG \to k \to 0$$

where the maps can be described quite explicitly. See Brown (1982, Section IV.2, Exercise 4c) for details.

Gruenberg (1970, Chapter 3) describes a related resolution derived from a presentation. That resolution requires a set of free generators for $R$ as a *subgroup* of $F$—which is more difficult to arrange. However, in some interesting cases it reduces to something more manageable (e.g. in the case of a cyclic group, it yields the efficient resolution in the previous section).

## 2.3  The bar resolution

Let $G$ be a group. For each $n \geq 0$, let $X_n$ be the free $kG$-module with basis $G^n$ (the $n$-fold cartesian product of $G$.) The basis elements are denoted

$$[x_1 | x_2 | \ldots | x_n]$$

where, for $n = 0$, the single basis element in $X_0$ is denoted $[\ ]$. The augmentation $\epsilon : X \to k$ is defined by $\epsilon([\ ]) = 1$. The differentials are defined by

$$d_n([x_1|x_2|\ldots|x_n]) = x_1[x_2|\ldots|x_n]$$
$$+ \sum_{i=1}^{n-1}(-1)^i[x_1|\ldots|x_{i-1}|x_ix_{i+1}|x_{i+2}\ldots|x_n]$$
$$+ (-1)^n[x_1|x_2|\ldots|x_{n-1}].$$

(The formula looks more symmetric if it is assumed that there is an $x_n$ pushed out on the right which acts trivially.) To see that the resulting complex is acyclic, define a contracting homotopy as follows. Define a $k$-homomorphism $\sigma : k \to X_0$ by $\sigma(1) = [\ ]$ and $k$-homomorphisms $s_n : X_n \to X_{n+1}$ by

$$s_n(x[x_1|x_2|\ldots|x_n]) = [x|x_1|x_2|\ldots|x_n].$$

Then $\epsilon \sigma = \mathrm{Id}$, $d_1 s_0 + \sigma \epsilon = \mathrm{Id}$, and $d_{n+1}s_n + s_{n-1}d_n = \mathrm{Id}$ for $n > 0$.

The bar resolution has the advantage that it is functorial in the group $G$. Arbitrary resolutions are of course functorial up to chain homotopy, but that is not quite the same. The bar resolution also provides a convenient interface where different approaches to group cohomology come together.

There is a useful modification of the bar resolution called the *normalized bar resolution* which is defined as follows. Consider the $kG$-submodule $T_n$ of $X_n$ generated by all $[x_1|x_2|\ldots|x_n]$ with at least one $x_i = 1$, and put $\overline{X}_n = X_n/T_n$. A simple calculation shows that $d_n(T_n) \subseteq T_{n-1}$, so $d_n$ induces $\overline{d}_n : \overline{X}_n \to \overline{X}_{n-1}$. Let $\overline{X}$ denote the resulting complex. We have $\overline{X}_0 = X_0$ so we may use the same augmentation $\epsilon : \overline{X}_0 \to k$. Similar reasoning applies to the contracting homotopy, so that $\epsilon : \overline{X} \to k$ is a free resolution of $k$.

The bar resolution provides the easiest way to identify low dimensional cohomology and homology groups.

Let $C^*(G, M) = \mathrm{Hom}_G(X, M)$ so, for $f \in C^n(G, M)$,

$$d^n f(x_1|\ldots|x_{n+1}) = x_1 f(x_2|\ldots|x_{n+1})$$
$$+ \sum_{i=1}^{n}(-1)^i f(x_1|\ldots|x_ix_{i+1}|\ldots|x_{n+1})$$
$$+ (-1)^{n+1} f(x_1|\ldots|x_n).$$

(If we use the normalized bar resolution, then dually we restrict attention to *normalized cocycles*, i.e. those which vanish when at least one argument $x_i = 1$.)

We can identify $C^0(G,M) = M$ where $d^0(m)(x) = (x-1)m$, so it is clear that $H^0(G,M) = M^G$ as expected. Similarly, $f \in C^1(G,M)$ is a 1-cocycle if and only if

$$xf(y) - f(xy) + f(x) = 0,$$

which amounts to saying that $f$ is a derivation. That derivation is a coboundary, $f = d^0 m$, if and only if $f(g) = (g-1)m$, and those derivations are called *inner* derivations. Thus

$$H^1(G,M) = \mathrm{Der}(G,M)/\mathrm{InDer}(G,M).$$

In particular, if $G$ acts trivially on $M$, then

$$H^1(G,M) = \mathrm{Hom}(G,M) = \mathrm{Hom}(G/G',M).$$

The case $n = 2$ is much more involved. $f \in C^2(G,M)$ is a 2-cocycle if and only if

$$xf(y|z) + f(x|yz) = f(xy|z) + f(x|y).$$

For each such cocycle we can put a group structure on the set $U = M \times G$ by defining

$$(a,x)(b,y) = (a + xb + f(x|y), xy).$$

In this context, the cocycle is often called a *factor set*. The associative law and other group properties are consequences of the cocycle condition. (The identity element is $(-f(1|1), 1)$.) Define a map $i : M \to U$ by $a \to (a - f(1|1), 1)$. It is not hard to check that $i$ is a monomorphism of groups. Moreover, $(a,x) \to x$ defines an epimorphism $p : U \to G$ with $\mathrm{Ker}\, p = \mathrm{Im}\, i$. Hence, $M$ may be identified with a normal subgroup of $U$ with $U/M \cong G$. This information may be summarized in a short exact sequence of groups

$$0 \to M \xrightarrow{i} U \xrightarrow{p} G \to 1.$$

If two cocycles $f_1$ and $f_2$ differ by a coboundary $d^1 g$, the corresponding $U_1$ and $U_2$ are isomorphic by an isomorphism which induces the identity on $M$ and $G$, i.e. we have a commutative diagram

$$\begin{array}{ccccc} M & \longrightarrow & U_1 & \longrightarrow & G \\ \Vert & & \downarrow & & \Vert \\ M & \longrightarrow & U_2 & \longrightarrow & G \end{array}$$

Such extensions are called *equivalent*. Conversely, given a group extension $M \xrightarrow{i} U \xrightarrow{p} G$ and a choice of coset representatives $u(x) \in U$ for the elements of $G$, since $i$ is a monomorphism, the formula

$$i(xm) = u(x) i(m) u(x)^{-1}$$

defines an action of $G$ on $M$. Also, this action does not depend on the specific choice of coset representatives since $M$ is abelian. Similarly, we may define a 2-cocycle $f$ by

$$i(f(x|y)) = u(x)u(y)u(xy)^{-1}.$$

Thus $H^2(G, M)$ may be identified with the set of equivalence classes of group extensions of $G$ by $M$. The latter even has a natural group structure called the Baer sum which corresponds to the operation in $H^2$. All this is tedious but fairly routine. See Brown (1982, Section IV.3) and Hilton and Stammbach (1971, Section VI.10) for details.

The group operation in $U$ is a bit simpler to deal with if we use *normalized* cochains. so that the cocycle associated with the extension will satisfy $f(1|x) = f(x|1) = 0$.

There are also interpretations for $H^3(G, M)$, but that theory is much more involved. See Eilenberg and Mac Lane (1947b), Mac Lane (1949), and Mac Lane (1963, Sections IV.8–9).

The bar resolution also allows one to interpret $H_1$ and $H_2$ in some cases. In particular, if $G$ acts trivially on $M$, we obtain

$$H_1(G, M) = G/G' \otimes_{\mathbf{Z}} M.$$

The cycle $[x] \otimes m$ determines a class on the left which corresponds to $xG' \otimes m$ on the right.

It is also easy to calculate $H_1(G, M)$ by using the long exact sequence in homology resulting from the sequence

$$0 \to I \to kG \to k \to 0.$$

That sequence shows that

$$\operatorname{Tor}_1^G(k, M) \cong \operatorname{Ker}\{I \otimes_G M \to M\}.$$

If $G$ acts trivially on $M$, then

$$\operatorname{Tor}_1^G(k, M) \cong I \otimes_G M \cong I/I^2 \otimes M.$$

It is easy to see that $x \to x - 1$ induces an isomorphism $G/G' \cong I/I^2$ for $k = \mathbf{Z}$, and $G/G' \otimes_{\mathbf{Z}} k \cong I/I^2$ in general.

A similar argument shows that for $n > 1$

$$\operatorname{Ext}_{kG}^{n-1}(I, M) \cong \operatorname{Ext}_{kG}^n(k, M).$$

It follows that the cohomology groups (with degrees shifted by 1) may be obtained as the right derived functors of $\operatorname{Der}(G, M) = \operatorname{Hom}_{kG}(I, M)$. That approach has certain advantages which were exploited by Barr and Rinehart (1966). (See also Rinehart (1969) and Ratcliffe (1980).) In particular, we can use this remark to calculate the cohomology of the free product of two groups. For, if $G = H * K$, it follows from the definition of free product (Hall 1959, Section 17.1) that

$$\mathrm{Der}(G, M) = \mathrm{Der}(H, M) \oplus \mathrm{Der}(K, M)$$

so

$$H^n(H * K, M) = H^n(H, M) \oplus H^n(K, M) \quad \text{for } n > 1.$$

This argument may also be generalized to yield a Mayer–Vietoris sequence for the free product with amalgamation. (See Exercise 4.1.2.)

**Exercise 2.3.1.** Determine what the formulas for the identity and inverses look like in the normalized case. Also, describe $i : M \to U$. Finally, show that if $f$ is a (possibly non-normalized) 2-cocycle, then $f' = f - d^1 g$, where $g(x) = f(1|x)$, is normalized.

**Exercise 2.3.2.** Let $M \xrightarrow{i} U \xrightarrow{p} G$ be a group extension. We say the extension splits if there is a group homomorphism $h : G \to U$ such that $ph = \mathrm{Id}$. Fix one such splitting, $h_0 : G \to U$. (This yields an action of $G$ on $M$ as above.) Any other splitting $h : G \to M$ is related to $h_0$ by $h(x) = i(g(x))h_0(x)$ where $g(x) \in M$.

(a) Show that $g$ is a 1-cocycle (i.e. a derivation). (The only difficulty is switching between multiplicative notation in $U$ and additive notation in $M$ without getting confused.)

(b) Show that $H^1(G, M)$ is in one-to-one correspondence with the set of equivalence classes of splittings $h : G \to U$ where two splittings $h_1$ and $h_2$ are considered equivalent if they differ by an inner automorphism by an element $i(m) \in i(M)$, i.e. $h_1(x) = i(m)h_2(x)i(m)^{-1}$ for $x \in G$.

**Exercise 2.3.3.** Show that $\mathrm{Hom}_k(I/I^2, k) \cong \mathrm{Ext}^1_{kG}(k, k)$.

## 2.4 Minimal resolutions

Let $k$ be a field. Then, for $G$ finite, $R = kG$ is an artinian ring. Let $J$ denote the Jacobson radical of $R$. Let

$$R = \oplus_i U_i$$

where the $U_i$ are indecomposable left ideals. In an artinian ring, any finitely generated projective module is a direct sum of such ideals, so the indecomposable projectives (up to isomorphism) are the same as the indecomposable left ideals of $R$. In addition, if $U$ is an indecomposable projective,

$U/JU$ is a simple $R$-module, and conversely for each simple $R$-module $V$ there is an indecomposable projective $U$ such that $U/JU \cong V$. Moreover, $U$ is unique in the following strong sense. Let $\epsilon : U \to V$ and $\epsilon' : U' \to V$ be epimorphisms from indecomposable projectives to $V$ with kernels $JU$ and $JU'$ respectively. Then there is an isomorphism $\phi : U \to U'$ such that

$$\begin{array}{ccc} U & & \\ & \searrow^{\epsilon} & \\ \phi \downarrow & V & \\ & \nearrow_{\epsilon'} & \\ U' & & \end{array}$$

commutes.

In general, the theory of injective modules parallels that described above, but is more complicated. However, for one class of artinian rings, the quasi-Frobenius rings, finitely generated projectives are injective and vice versa. (That is almost the definition of 'quasi-Frobenius'.) Thus, indecomposable projectives are also indecomposable injectives, and dual versions of the above assertions hold true in that each simple module is associated with an indecomposable injective (projective) as its unique simple submodule. A group algebra over a field is quasi-Frobenius, and moreover for group algebras the unique simple quotient module and the unique simple submodule are the same. See Curtis and Reiner (1981, Introduction, Section 6) or Benson (1984, Section 1.4) for basic definitions and details.

Using the indecomposable projectives, we may construct a projective resolution of a finitely generated $R$-module $M$ as follows. $M/JM$ is a semisimple module so it is a direct sum of simple modules. For each simple constituent $V_i$, choose an indecomposable projective $U_i$ and an epimorphism $U_i \to V_i$. Add these up to obtain an epimorphism $g : U \to M/JM$ where $U = \bigoplus_i U_i$. Note that $g$ induces $\bar{g} : U/JU \cong M/JM$. Since $U$ is projective, there is an $\epsilon : U \to M$ such that the diagram

$$\begin{array}{ccc} U & \longrightarrow & U/JU \\ \epsilon \downarrow & & \downarrow \bar{g} \\ M & \longrightarrow & M/JM \longrightarrow 0 \end{array}$$

commutes. It follows that $M = \operatorname{Im}\epsilon + JM$ so by Nakayama's Lemma, we conclude that $\epsilon$ is an epimorphism. Note that $\operatorname{Ker}\epsilon \subseteq JU$. Reasoning as in the simple case above, we can see that these conditions characterize $U \to M$ up to isomorphism (over $M$). We call $U \to M$ a *projective cover*.

Let $P_0 = U$, and apply the same argument to $\operatorname{Ker}\epsilon$ to obtain a projective cover $P_1 \to \operatorname{Ker}\epsilon$. Continuing in this way, we obtain a resolution

$$\cdots \to P_n \xrightarrow{d_n} P_{n-1} \to \cdots \to P_1 \xrightarrow{d_1} P_0 \xrightarrow{\epsilon} M \to 0$$

such that $\operatorname{Ker}\epsilon = \operatorname{Im} d_1 \subseteq JP_0$, and for $n > 0$, $\operatorname{Ker} d_n = \operatorname{Im} d_{n+1} \subseteq JP_n$. Such a resolution is called a *minimal resolution*. It is not hard to see that the special resolution of $k$ for cyclic groups is minimal if $k$ is a field.

Minimal resolutions are unique up to isomorphism of complexes. Indeed, suppose that $\epsilon : P \to M$ is minimal, and $\eta : X \to M$ is any projective resolution (with each $X_n$ finitely generated.) Then there exist $F : P \to X$ and $G : X \to P$ such that $FG$ and $GF$ are each chain homotopic to the identity. However, we may modify $G$ so that $G_n F_n = \operatorname{Id}$ for each $n \geq 0$. To see this note that, for $n = 0$, the image of $T_n = G_n F_n - \operatorname{Id}$ is contained in $JP_n$, and since $J$ is nilpotent, it is not hard to see that $G_n F_n = \operatorname{Id} + T_n$ is invertible. Replacing $G_n$ by $(\operatorname{Id} + T_n)^{-1} G_n$ yields $G_n F_n = \operatorname{Id}$. This process may be continued inductively. It follows that, as complexes,

$$X \cong P \oplus Y, \qquad (2.1)$$

where $Y$ is a projective resolution of 0. If $X \to M$ is also minimal, the same argument applies to $FG$, so $X \cong P$ as complexes over $M$.

The advantage of using a minimal resolution is that if $W$ is any *simple* module, then the differentials in the complexes $\operatorname{Hom}_R(P, W)$ and $P \otimes_R W$ are trivial. For example, if $f \in \operatorname{Hom}_R(P_n, W)$, then $f(JP_n) = Jf(P_n) \subseteq JW = 0$ so $d^n f = f d_{n+1} = 0$, and a similar argument works for tensor products. Hence,

$$\operatorname{Tor}_n^R(M, W) = P_n \otimes_R W$$
$$\operatorname{Ext}_R^n(M, W) = \operatorname{Hom}_R(P_n, W).$$

Each $P_n$ is a sum of indecomposable projectives $U_i$ with $U_i/JU_i = V_i$ simple, and for each such summand $\operatorname{Hom}_R(U_i, W) \cong \operatorname{Hom}_R(V_i, W)$. By Schur's Lemma, the latter is trivial unless $V_i \cong W$ in which case it is a division algebra over $k$, usually just $k$ itself. Similar remarks apply to the calculation of Tor.

To exploit minimal resolutions, one must know quite a lot about the modular representation theory of the group $G$. Benson and Carlson have derived some very interesting results by this approach building on earlier work of Alperin.

**Exercise 2.4.1.** Let $G$ be a finite group and $k$ a field. Let $M$ be a finitely generated $kG$-module. The *socle* $S(M)$ is defined to be the maximal semi-simple submodule of $M$. It may also be characterized as $\{x \in M \mid Jx = 0\}$. Show that there is an imbedding $i : M \to U$ in a finitely generated injective module $U$ such that $i(S(M)) = S(U)$. Show that $U$ is unique up to an isomorphism over $M$. $M \to U$ is called an *injective hull*.

**Exercise 2.4.2.** Let $G$ be a finite group and $k$ a field.

(a) Suppose $M$ is a finitely generated $kG$-module without non-trivial projective summands. Show that if $U \to M$ is a projective cover with kernel $K$, then $K \to U$ is an injective hull. Conversely, let $K$ be finitely generated without non-trivial projective (i.e. injective) summands, and suppose $i : K \to U$ is an injective hull. Show that $U \to \text{Coker } i$ is a projective cover. (Curtis and Reiner 1981, Introduction, Theorem 6.31.)

(b) Suppose $M$ is a finitely generated $kG$-module without non-trivial projective summands. Show that the same is true for each $\text{Im } d_i$ in its minimal resolution.

(c) Using (a) and (b), show that the minimal resolution for any finitely generated module $M$ without non-trivial projective summands is completely determined by any $\text{Im } d_i$.

If $G$ is a $p$-group, the situation is simpler than in general. The reason is that for $k$ a field of characteristic $p > 0$, there is only one simple $kG$-module, namely $k$ itself with trivial action. (To prove this, show by induction on $|G|$ that $M^G \neq \{0\}$. The case $|G| = p$ follows by noting that, for $g$ a generator of $G$ and any $x \in M$, there is a least power $i$ such that $(g-1)^i x = 0$, whence $(g-1)^{i-1} x \in M^G$.) It follows that if $G$ is a $p$-group, the Jacobson radical is the augmentation ideal, the only indecomposable projective is $kG$ itself, and every projective (injective) is free.

We shall use some of these ideas in a fairly trivial way to prove a famous theorem of Golod and Safarevec (1964) which was used to settle negatively an important conjecture in algebraic number theory about class field towers. Their theorem relates $\dim H^2(G, \mathbf{F}_p)$ and $\dim H^1(G, \mathbf{F}_p)$ which are important invariants in the theory of presentations of finite $p$-groups. We give a version of the theorem due to Roquette (1967).

**Theorem 2.4.1.** *Let $G$ be a finite $p$-group, and $k = \mathbf{F}_p$. Let*

$$d = \dim_k H^1(G, k)$$
$$r = \dim_k H^2(G, k).$$

*Then*

$$r > d^2/4.$$

**Proof.** Let $R = kG$. As mentioned above, the only indecomposable projective is $R$ itself, and its Jacobson radical is the augmentation ideal $I$. Hence, the minimal resolution of $k$ starts off

$$0 \to I \to R \to k \to 0$$
$$Q \xrightarrow{g} P \xrightarrow{f} I \to 0$$

where $P = P_1$ has rank $d$ over $kG$ and $Q = P_2$ has rank $r$. Let $Q_j = g^{-1}(I^j P)$ for $j = 0, 1, \ldots$ so that we have exact sequences

$$Q_j \to I^j P \to I^{j+1} \to 0.$$

Note that $Q_0 = Q$ and, since $\operatorname{Im} g = \operatorname{Ker} f \subseteq IP$, $Q_1 = Q$. Tensoring with $k$ yields

$$Q_j/IQ_j \to I^j P/I^{j+1} P \to I^{j+1}/I^{j+2} \to 0,$$

and it is easy to see that $IQ_j \subseteq Q_{j+1}$ and

$$0 \to Q_j/Q_{j+1} \to I^j P/I^{j+1} P \to I^{j+1}/I^{j+2} \to 0$$

is exact. Let $d_j = \dim I^j/I^{j+1}$ and $e_j = \dim Q_j/Q_{j+1}$. Note that since $I^0 = kG$, $d_0 = 1$. Since $P/IP \cong I/I^2$, $d_1 = d$. Since $P$ is free of rank $d$, the above sequence tells us that

$$e_j + d_{j+1} = dd_j \qquad \text{for } j = 0, 1, \ldots. \tag{2.2}$$

Iterating the above inclusion shows that $I^j Q \subseteq Q_{j+1}$ so that

$$\dim Q/I^j Q \geq \dim Q/Q_{j+1}.$$

So by adding up dimensions of the factors, we have

$$r(d_0 + \cdots + d_{j-1}) \geq e_0 + \cdots + e_j \qquad \text{for } j \geq 1. \tag{2.3}$$

Form the Poincaré series

$$e(t) = \sum_j e_j t^j$$
$$d(t) = \sum_j d_j t^j.$$

Note that these are in fact polynomials since $I$, being the Jacobson radical of $kG$, is nilpotent and the $d$'s and $e$'s vanish for large $j$. From the dimension equality (2.2), using $d_0 = 1$, we have

$$e(t) + (d(t) - 1)/t = dd(t) \qquad \text{or}$$
$$e(t) = ((dt - 1)d(t) + 1)/t.$$

The coefficient of $t^j$ in $e(t)/(1-t)$ is $e_0 + \cdots + e_j$, and similarly for $d(t)$. Since $e_0 = 0$, it follows from (2.3) that

$$rtd(t)/(1-t) \geq e(t)/(1-t)$$

for all $0 < t < 1$. Hence,

$$rtd(t) \geq ((dt-1)d(t)+1)/t \quad \text{or}$$
$$(rt^2 - dt + 1)d(t) \geq 1$$

for $0 < t < 1$. Since $d(t) > 0$ in that range, the quadratic coefficient must be positive there. However, its minimum occurs at $t = d/2r$ which also lies in the interval. For, if $q = \dim kG$, consideration of the sequence

$$Q \to P \to I \to 0$$

shows that

$$rq + q - 1 \geq dq \quad \text{or}$$
$$r \geq d - 1 + 1/q \quad \text{or}$$
$$r \geq d$$

since both are integers. Hence, the discriminant must be negative, i.e.

$$d^2 - 4r < 0$$

which gives the desired inequality. $\square$

**Exercise 2.4.3.** Let $G$ be a non-trivial finite $p$-group, and let $k$ be a field of characteristic $p$ with $G$ acting trivially.
 (a) Show that $H^n(G, k) \neq 0$ for $n \geq 0$. Hint: Use a minimal resolution.
 (b) Let $M$ be a finitely generated $kG$-module. Show that $M$ is projective (i.e. free) if and only if $H^n(G, M) = 0$ for at least one $n > 0$. Hint: Develop the theory of minimal injective resolutions. Alternatvely, relate $\operatorname{Ext}^*_{kG}(M^*, k)$ to $\operatorname{Ext}^*_{kG}(k, M)$.

## 2.5  Building new resolutions from old resolutions

We shall assume henceforth that the base ring $k$ is a principal ideal domain, but essentially everything we do will go through for Dedekind domains with minor modifications. The most important examples are $k = \mathbf{Z}$ or $k$ a field. Let $A$ and $B$ be $k$-free chain complexes, i.e. assume each homogeneous component is $k$-free. Make $A \otimes B$ into a double complex as usual by defining

$$d'(a \otimes b) = da \otimes b$$
$$d''(a \otimes b) = (-1)^{\deg a} a \otimes db.$$

Then $d = d' + d''$ is the differential of the associated single complex. The Künneth formula tells us that there is an exact sequence

$$0 \to H_*(A) \otimes H_*(B) \to H_*(A \otimes B) \to \operatorname{Tor}_1^k(H_*(A), H_*(B)) \to 0$$

from which we may calculate the homology of the associated single complex. The left hand map is defined as follows: if $a$ and $b$ are cycles, then

$$cls(a) \otimes cls(b) \mapsto cls(a \otimes b).$$

The right hand map is much trickier to define and it lowers the total degree by 1. Moreover, the right hand map splits, but not functorially. (See Mac Lane (1963, Section V.10).)

Suppose now that $G$ and $H$ are groups and $\epsilon : X \to k$ and $\eta : Y \to k$ are projective resolutions for $G$ and $H$ respectively. One way to assert the acyclicity of the resolutions is to say that $\epsilon_* : H_*(X) \cong k$ and $\eta_* : H_*(Y) \cong k$. Since $kG$ is $k$-free, it follows that $X$ and $Y$ are $k$-free complexes. Hence, the Künneth formula and the fact that $\operatorname{Tor}_1^k(k,-) = 0$ tells us that $H_*(X) \otimes H_*(Y) \cong H_*(X \otimes Y)$, from which we can deduce that $\epsilon \otimes \eta$ induces an isomorphism of the latter homology group with $k$ so that $X \otimes Y$ is an acyclic complex over $k$. Also, since $k(G \times H) \cong kG \otimes kH$, it follows by considering direct summands that $X \otimes Y \to k$ is a $k(G \times H)$-projective resolution. (Note that this argument would have worked for projective resolutions of any $k$-free modules, so it would always work if $k$ is a field, but not necessarily otherwise.)

It is clearly appropriate to use this construction to investigate the homology or cohomology of direct products. We illustrate some typical arguments for cohomology. Let $M$ be a $kG$-module and $N$ a $kH$-module. We shall *assume* that the resolution $X$ (similarly $Y$) has been chosen to be *finitely generated as a $kG$-module ($kH$-module) in each degree*. This is always possible for finite groups and is usually possible for other interesting classes of groups. In this case, the map of complexes

$$\operatorname{Hom}_G(X, M) \otimes \operatorname{Hom}_H(Y, N) \to \operatorname{Hom}_{G \times H}(X \otimes Y, M \otimes N)$$

defined by

$$f \otimes g \mapsto f \times g$$

where

$$(f \times g)(x \otimes y) = f(x) \otimes g(y)$$

is an isomorphism. It follows that

$$H^*(G \times H, M \otimes N) = H^*(\operatorname{Hom}_G(X, M) \otimes \operatorname{Hom}_H(Y, N)).$$

*Assume further* that $M$ or $N$ is $k$-free so that one of the complexes in the product on the right is $k$-free. Then, we may apply the Künneth formula again to conclude that there is a split exact sequence

$$0 \to H^*(G,M) \otimes H^*(H,N) \to H^*(G \times H, M \otimes N)$$
$$\to \operatorname{Tor}_1^k(H^*(G,M), H^*(H,N)) \to 0.$$

Because we are dealing with cohomology (and using the formal relation $C^n = C_{-n}$), the map on the right *raises* degree by 1. In particular, if $k$ is a field, we obtain an isomorphism

$$H^*(G,M) \otimes H^*(H,N) \cong H^*(G \times H, M \otimes N).$$

These remarks suffice to calculate the cohomology of abelian groups. Suppose, for example, that $P$ is cyclic of prime order $p$ and $G = P^d$ is an elementary abelian $p$-group. If $k$ is a field, we conclude immediately that

$$H^*(G,k) \cong H^*(P,k)^{\otimes d} = \overbrace{H^*(P,k) \otimes H^*(P,k) \otimes \cdots \otimes H^*(P,k)}^{d \text{ times}}$$

and, since we calculated $H^*(P,k)$ in Section 2.1, we can calculate the cohomology of the direct product $P^d$.

If $k$ is not a field (but still a PID), the situation is more complicated. Let $G = P^d$ as above, and let $X \to k$ be the $kP$-free resolution introduced earlier in Section 2.1. Then $H^*(G,k)$ is determined by calculating the cohomology of the cochain complex $\operatorname{Hom}_{kP}(X,k)^{\otimes d}$. However, for each $n$, $\operatorname{Hom}_P(X_n, k) \cong k$ and, since $P$ acts trivially on $k$, the differentials alternate between the zero map and the map which multiplies by the order $p$ of $P$. That gives us a very explicit complex in which it is possible to make detailed calculations. Of course, we could just use the Künneth Theorem as above, but the approach just outlined gives more precise information.

The above method works just as well for any finitely generated abelian group. The case $\mathbf{Z}^d$ is specially simple, and it is left for the reader to investigate as an exercise. For finite abelian groups, the answer is similar to that for $P^d$ except that if the cyclic factors have different orders, the torsion terms can become rather complicated.

The next case to consider is that of *semi-direct products*. Suppose then that $G$ and $H$ are groups and a homomorphism $\alpha : G \to \operatorname{Aut}(H)$ is specified. Write $g(h) = \alpha(g)(h)$ for $g \in G, h \in H$. Then the semi-direct product $U = H \rtimes G$ consists of all pairs $(h,g)$ with the law of composition

$$(h_1, g_1)(h_2, g_2) = (h_1 g_1(h_2), g_1 g_2).$$

If $H$ is abelian and written additively, it is sometimes useful to think of $H \rtimes G$ as consisting of all 'matrices'

$$\begin{bmatrix} g & h \\ 0 & 1 \end{bmatrix}$$

where a suitable interpretation of matrix multiplication gives the proper operation. If $H = V$ is a vector space and $G$ is a subgroup of $Gl(V)$, then we have a bona fide matrix representation of this form.

Let $W \to k$ be a $kG$-projective resolution, and suppose $X \to k$ is a $kH$-projective resolution *on which $G$ acts* in a manner consistent with its action on $H$, i.e. the augmentation and differential are $G$-maps and $g(hx) = g(h)gx$ for $g \in G, h \in H, x \in X$. For example, if $X$ is the bar resolution for $H$, we may define such a $G$-action by

$$g(h[h_1|h_2|\ldots|h_n]) = g(h)[g(h_1)|g(h_2)|\ldots|g(h_n)].$$

It is easy to check that the differential and augmentation in the bar resolution are $G$-maps.

Given $W$ and $X$, as above, we make $W \otimes X$ into a $H \rtimes G$-complex by defining

$$(h, g)(w \otimes x) = gw \otimes h(gx),$$

and as above, it is an acyclic complex over $k$.

**Proposition 2.5.1.** *If $U$ is a projective $kG$-module and $V$ is a projective $kH$-module on which $G$ acts as above, then $U \otimes V$ is a projective $k(H \rtimes G)$-module.*

**Proof.** We have a natural equivalence of functors on the category of $H \rtimes G$-modules $M$

$$\mathrm{Hom}_{H \rtimes G}(U \otimes V, M) \cong \mathrm{Hom}_G(U, \mathrm{Hom}_H(V, M))$$

defined by $f \mapsto F$ where $f(u \otimes v) = F(u)(v)$. It is left to the reader to check the required details. Since $V$ is $kH$-projective, the functor $\mathrm{Hom}_H(V, -)$ is exact, and since $U$ is $kG$-projective the same is true of $\mathrm{Hom}_G(U, -)$. Thus $\mathrm{Hom}_{H \rtimes G}(U \otimes V, -)$ is exact since it is the composite of two exact functors; hence $U \otimes V$ is $k(H \rtimes G)$-projective. □

It follows that $W \otimes X \to k$ is a $k(H \rtimes G)$-projective resolution and we even have the isomorphism of complexes

$$\mathrm{Hom}_{H \rtimes G}(W \otimes X, M) \cong \mathrm{Hom}_G(W, \mathrm{Hom}_H(X, M)).$$

This isomorphism is not nearly as useful as the corresponding isomorphism for direct products since there are no general theorems which help us compute the cohomology on the right. However, suppose $\mathrm{Hom}_H(X, M)$ has a trivial differential. That might be the case, for example, if $k$ were a field and $X \to k$ were constructed starting with minimal resolutions. Then $H^*(\mathrm{Hom}_H(X, M)) = \mathrm{Hom}_H(X, M)$, so we can conclude

$$H^*(H \rtimes G, M) \cong H^*(G, \mathrm{Hom}_H(X, M)) = H^*(G, H^*(H, M)).$$

## Explicit resolutions

In fact, one can rarely find a minimal $H$-resolution on which $G$ acts, but sometimes one can build up such an $H$-resolution by piecing together minimal resolutions. The only important application of this principle is to wreath products, which we shall investigate in due course.

**Exercise 2.5.1.** Let $G$ be a group, and take $M = k = \mathbf{F}_p$. Define

$$h_G(t) = \sum_{n=0}^{\infty} \dim_k H^n(G,k) \, t^n.$$

$h_G(t)$ is sometimes called the *Poincaré series* of $G$. Show that for $G$ an elementary abelian $p$-group of rank $d$, $h_G(t) = 1/(1-t)^d$.

# 3
# Products in cohomology

$H^*(G,k)$ may be endowed with a multiplicative structure which turns it into a commutative graded ring. (Here, 'commutative graded' means that the product satisfies the graded commutation rule $\alpha\beta = (-1)^{pq}\beta\alpha$ for $\alpha \in H^p(G,k)$, $\beta \in H^q(G,k)$.) Much of this monograph will be concerned with what is known about the structure of this ring and how it is related to the structure of $G$. At present, unfortunately, there is relatively little such knowledge.

The product—often called the *cup* product—was originally defined by means of the bar resolution by defining a product $fg$ in $C^*(G,k)$ which made it a *differential graded algebra.*. That means that the coboundary morphism $d: C^*(G,k) \to C^*(G,k)$ is a *derivation*, i.e. it satisfies

$$d(fg) = (df)g + (-1)^p f dg \qquad \text{for } f \in C^p(G,k), g \in C^q(G,k).$$

The product in this algebra is not exactly commutative but only commutative (in the graded sense) modulo coboundaries.

For our purposes, it will be important to introduce the product structure in a somewhat more general context. First, we need products for coefficient modules other than the ring $k$, and we also want to be able compute products at the cochain level for resolutions other than the bar resolution.

## 3.1 Definitions

Given $G$-modules $A$ and $B$, the *cup product* will be a collection of homomorphisms
$$H^r(G,A) \otimes H^s(G,B) \to H^{r+s}(G, A \otimes B).$$

(In general, we shall also use the term 'pairing' for a homomorphism of a tensor product into a third group.) If in addition we are given a $G$-homomorphism ($G$-pairing) $A \otimes B \to C$, we may compose to produce a collection of pairings
$$H^r(G,A) \otimes H^s(G,B) \to H^{r+s}(G,C)$$

called the *cup product* relative to $A \otimes B \to C$. In particular, if $A = B = C$ is a ring and the multiplication map $A \otimes A \to A$ is a $kG$-morphism, then the cup product makes $H^*(G, A)$ into a graded ring. More generally, if $A$ is a $kG$-ring as above and $M$ is both a $G$-module and an $A$-module such that the action $A \otimes M \to M$ is a $G$-morphism, then $H^*(G, M)$ is a graded module over $H^*(G, A)$. The most interesting case is, as mentioned above, $A = k$ with $G$ acting trivially.

The cup product is defined in two stages. First, let $G$ and $H$ be groups, and let $A$ be a $kG$-module and $B$ a $kH$-module. Let $X \to k$ be a $kG$-projective resolution and $Y \to k$ a $kH$-projective resolution. Define the *cross* product

$$\operatorname{Hom}_G(X, A) \otimes \operatorname{Hom}_H(Y, B) \xrightarrow{\times} \operatorname{Hom}_{G \times H}(X \otimes Y, A \otimes B)$$

by

$$(f \times g)(x \otimes y) = f(x) \otimes g(y)$$

for $f \in \operatorname{Hom}_G(X, A)$, $g \in \operatorname{Hom}_H(Y, B)$. This induces a map on the cohomology of these complexes which, when composed with the Künneth map, induces a homomorphism

$$H^*(G, A) \otimes H^*(H, B) \to H^*(G \times H, A \otimes B)$$

which preserves (total) degree. If $\alpha \in H^r(G, A), \beta \in H^s(H, B)$ we denote the image of $\alpha \otimes \beta$ in $H^{r+s}(G \times H, A \otimes B)$ by $\alpha \times \beta$.

Suppose that $G$ is a group and $A$ and $B$ are $kG$-modules. Let $\Delta : G \to G \times G$ denote the diagonal homomorphism defined by $\Delta(x) = x \times x$, and let $G$ act on $A \otimes B$ through $\Delta$. Then by functorality there is an induced homomorphism

$$\Delta^* : H^*(G \times G, A \otimes B) \to H^*(G, A \otimes B).$$

(Remember that cohomology is contravariant in $G$.) Define the cup product as the composition of $\Delta^*$ with the cross product. Thus for $\alpha \in H^r(G, A)$ and $\beta \in H^s(G, B)$, $\alpha\beta \in H^{r+s}(G, A \otimes B)$ is given by

$$\alpha\beta = \Delta^*(\alpha \times \beta).$$

The cup product is often denoted $\alpha \cup \beta$, but it is simpler to denote it by ordinary product notation.

In order to derive the formal properties of the cup product, we must state the corresponding formal properties for the cross product. The proofs are perfectly clear from the definitions, and the reader is advised to work some of them out.

(a) Let $\phi : G' \to G$ be a group homomorphism. If $A$ is a $G$-module and $A'$ is a $G'$-module, then $A$ may be viewed as a $G'$-module through $\phi$.

## Definitions

Suppose $f : A \to A'$ is a $G'$-module homomorphism, i.e. $f(\phi(x')a) = x'f(a)$ for $x' \in G'$ and $a \in A$. Then there is an induced homomorphism

$$f^*\phi^* = (\phi, f)^* : H^*(G, A) \to H^*(G', A').$$

Suppose similarly that there is such a pair $\psi : H' \to H$, $g : B \to B'$. Then there is a commutative diagram

$$\begin{array}{ccc} H^*(G, A) \otimes H^*(H, B) & \xrightarrow{\times} & H^*(G \times H, A \otimes B) \\ {\scriptstyle (\phi,f)^* \otimes (\psi,g)^*} \downarrow & & \downarrow {\scriptstyle (\phi\times\psi, f\otimes g)^*} \\ H^*(G', A') \otimes H^*(H', B') & \xrightarrow{\times} & H^*(G' \times H', A' \otimes B') \end{array}$$

(b) If we have three groups $G, H,$ and $K$ and corresponding modules $A, B,$ and $C$, then, assuming we identify

$$(A \otimes B) \otimes C = A \otimes (B \otimes C) = A \otimes B \otimes C,$$

we have

$$(\alpha \times \beta) \times \gamma = \alpha \times (\beta \times \gamma).$$

(c) If 1 denotes the trivial group, then $H^*(1, k) = H^0(1, k) = k$, and we may identify $k \otimes A = A$. Then for $1 \in k$, $\alpha \in H^*(G, A)$ we have $1 \times \alpha = \alpha$. With similar conventions, $\alpha \times 1 = \alpha$.

(d) Let $A$ be a $G$-module and $B$ an $H$-module. Let $\tau : H \times G \to G \times H$ and $t : A \otimes B \to B \otimes A$ be the morphisms which interchange factors. Then for $\alpha \in H^r(G, A)$ and $\beta \in H^s(G, B)$, we have

$$(\tau, t)^*(\alpha \times \beta) = (-1)^{rs} \beta \times \alpha.$$

Another way to say this is that the diagram

$$\begin{array}{ccc} H^*(G, A) \otimes H^*(H, B) & \xrightarrow{\times} & H^*(G \times H, A \otimes B) \\ T \downarrow & & \downarrow {\scriptstyle (\tau,t)^*} \\ H^*(H, B) \otimes H^*(G, A) & \xrightarrow{\times} & H^*(H \times G, B \otimes A) \end{array}$$

commutes where the map $T$ on the left interchanges factors and multiplies by the sign $(-1)^{rs}$. (That is what we would expect from the sign convention for graded objects.) Note that $T$ is induced by the interchange map for complexes $X \otimes Y \to Y \otimes X$. The sign must be included for $T$ to commute with the differentials.

The corresponding facts for the cup product follow readily.

(a) Let $\phi : G' \to G$ be a homomorphism of groups and suppose $f : A \to A'$ and $g : B \to B'$ are consistent module homomorphisms as above. Then $(\phi, f \otimes g)^*(\alpha\beta) = (\phi, f)^*(\alpha)(\phi, g)^*(\beta)$. This follows easily using the commutative diagram of group homomorphisms

$$\begin{array}{ccc} G' & \xrightarrow{\Delta'} & G' \times G' \\ \phi \downarrow & & \downarrow \phi \times \phi \\ G & \xrightarrow{\Delta} & G \times G \end{array}$$

(b) Let $A, B,$ and $C$ be $G$-modules. Then for $\alpha \in H^*(G, A)$, $\beta \in H^*(G, B)$, and $\gamma \in H^*(G, C)$, we have $(\alpha\beta)\gamma = \alpha(\beta\gamma)$.

(c) Let 1 be the identity element of $k = H^0(G, k)$. Then if we identify $k \otimes A = A$ and $A \otimes k = A$, we have $1\alpha = \alpha 1 = \alpha$ for $\alpha \in H^*(G, A)$. To prove this use the assertions (a) and (c) above for the cross product and the fact that if $\pi : G \to 1$ is the trivial group homomorphism then we have a commutative diagram

$$\begin{array}{ccc} G & \xrightarrow{\Delta} & G \times G \\ \mathrm{Id} \downarrow & & \downarrow \pi \times \mathrm{Id} \\ G & \xrightarrow{=} & 1 \times G \end{array}$$

Also use the fact that $\pi^*(1) = 1 \in H^0(G, k)$.

(d) Let $A$ and $B$ be $G$-modules, and let $t : A \otimes B \to B \otimes A$ interchange factors. ($t$ is a $G$-morphism.) Then for $\alpha \in H^r(G, A)$, $\beta \in H^s(G, B)$,

$$t^*(\alpha\beta) = (-1)^{rs}\beta\alpha.$$

To prove this it suffices to note that $\tau\Delta = \Delta$.

If $A$ is a $kG$-ring, then we may conclude from the above statements that $H^*(G, A)$ is an associative graded ring with identity $1 \in H^0(G, A) = A^G$ (the image of $1 \in k$). Moreover, if $M$ is an $A$-module which is also a $kG$-module with consistent action, then we may conclude that $H^*(G, M)$ is a graded $H^*(G, A)$-module—with 1 acting as the identity. Finally, if $A$ is a commutative $kG$-ring, it follows that $H^*(G, A)$ is a *commutative graded ring*, i.e. $\beta\alpha = (-1)^{rs}\alpha\beta$ for $\alpha$ homogeneous of degree $r$ and $\beta$ homogeneous of degree $s$. Note also that by functorality (property (a)), induced maps in cohomology are ring homomorphisms (or module homomorphisms in the module case).

## 3.2 Computations

It is often useful to describe the cup product explicitly at the level of cochains. For this purpose we need an explicit description of the map $\Delta^* : H^*(G \times G, A \otimes B) \to H^*(G, A \otimes B)$.

Let $X \to k$ be a projective $G$-resolution. Then on the left we would be using the $G \times G$-projective resolution $X \otimes X \to k$. Hence, to describe the map $\Delta^*$ we need a map of *complexes* $D : X \to X \otimes X$ which is consistent with augmentations and the homomorphism $\Delta : G \to G \times G$, i.e. $D(gx) = \Delta(g)D(x)$. Such a map is unique up to chain homotopy. We give examples of such *diagonal* maps $D$ for interesting complexes.

*The bar resolution*: Let $B \to k$ be the bar resolution as defined earlier. Then it is not hard to check that the following map—called the Alexander-Whitney map—is an appropriate diagonal map:

$$D[x_1|x_2|\ldots|x_n] = \sum_{r=1}^{n} [x_1|\ldots|x_r] \otimes (x_1\ldots x_r)[x_{r+1}|\ldots|x_n].$$

If $f \in C^r(G, A)$ and $g \in C^s(G, B)$ are cochains for the bar resolution, then at this level the cup product is given by

$$(f \cup g)(x_1|x_2|\ldots|x_{r+s}) = f(x_1|\ldots|x_r) \otimes (x_1\ldots x_r)g(x_{r+1}|\ldots|x_{r+s}).$$

At the cochain level using the bar resolution, the cup product is associative with identity, but the commutativity formulas mentioned above only hold up to cochain homotopy.

*The special resolution for cyclic groups*: Let $G = \langle g \mid g^m = 1 \rangle$ and let $X_n = kGx_n$, $\epsilon(x_0) = 1$, $d_n(x_n) = (g-1)x_{n-1}$ for $n$ odd and $d_n(x_n) = Tx_{n-1}$ for $n$ even, $n > 0$. Then the following map is a diagonal map (see Cartan and Eilenberg 1956, Section XII.7):

$D = \sum D_{r,s}$ where $D_{r,s} : X_{r+s} \to X_r \otimes X_s$, and

$$D_{r,s}(x_{r+s}) = x_r \otimes x_s \qquad r \text{ even}$$

$$D_{r,s}(x_{r+s}) = x_r \otimes gx_s \qquad r \text{ odd}, s \text{ even}$$

$$D_{r,s}(x_{r+s}) = \sum_{0 \leq i < j < m} g^i x_r \otimes g^j x_s \qquad r, s \text{ odd}.$$

A tedious but fairly routine calculation shows that this map commutes with the relevant differentials in $X$ and $X \otimes X$. Using this diagonal, we may calculate products for a cyclic group $G$. Thus if $A$ and $B$ are $G$-modules and $a \in A$ and $b \in B$ represent cohomology classes in degrees $r$ and $s$ respectively, the cup product of these classes is represented by

$$a \otimes b \qquad r \text{ or } s \text{ even}$$

$$\sum_{0 \leq i < j < m} g^i a \otimes g^j b \qquad r, s \text{ odd}.$$

(We use here that $gb = b$ if $b \in B^G$ represents a cohomology class of *even* degree.) In particular, if $A$ is a ring on which $G$ acts trivially, it follows that the ring structure is given by

$$\begin{array}{ll} ab & r \text{ or } s \text{ even} \\ \dfrac{m(m-1)}{2} ab & r, s \text{ odd.} \end{array}$$

Let $k$ be a field of characteristic $p$ dividing $|G|$ and suppose $p^c$ is the exact power of $p$ dividing $|G|$. Then it follows that

$$H^*(G, k) = k[\eta, \xi \mid \deg \eta = 1, \deg \xi = 2, \eta^2 = 0]$$

if $p$ is odd or if $p = 2$ and $c > 1$, and

$$H^*(G, k) = k[\eta \mid \deg \eta = 1]$$

if $p = 2$ and $c = 1$. Also, if $M$ is any $kG$-module, then multiplication by $\xi$ is an isomorphism $H^q(G, M) \to H^{q+2}(G, M)$ for $q > 0$ in general, and multiplication by $\eta$ is an isomorphism $H^q(G, M) \to H^{q+1}(G, M)$ for $q > 0$ if $p = 2$ and $c = 1$. We only obtain an epimorphism in general for $q = 0$.

Note that $H^1(G, k) = \mathrm{Hom}_{\mathbb{Z}}(G, k)$ since $G$ acts trivially on $k$, and $\eta$ may be chosen to be the homomorphism characterized by $\eta(g) = 1$.

## 3.3   Connecting homomorphisms and Bocksteins

Let $G$ be a group and let $0 \to A' \to A \to A'' \to 0$ be a short exact sequence of $kG$-modules. Then we know from homological algebra that there is a natural *connecting homomorphism* $\delta : H^*(G, A'') \to H^*(G, A')$ which raises degree by 1 and such that the sequence

$$\cdots \to H^n(G, A') \to H^n(G, A) \to H^n(G, A'') \xrightarrow{\delta} H^{n+1}(G, A') \to \cdots$$

is exact. Suppose that $G$ and $H$ are groups, $A$ is a $kG$-module, and $0 \to B' \to B \to B'' \to 0$ is an exact sequence of $kH$-modules. Let $C = \mathrm{Im}\{A \otimes B' \to A \otimes B\}$ and let $j : A \otimes B' \to C$ be the accompanying epimorphism. Then $0 \to C \to A \otimes B \to A \otimes B'' \to 0$ is an exact sequence of $G \times H$-modules.

**Proposition 3.3.1.** *With the notation as above, if $\alpha \in H^r(G, A)$ and $\beta'' \in H^s(H, B'')$, then*

$$\delta(\alpha \times \beta'') = (-1)^r j^*(\alpha \times \delta(\beta'')).$$

*Connecting homomorphisms and Bocksteins* 27

With suitable reversal of roles for $G$ and $H$, and with a short exact sequence of $kG$-modules,
$$\delta(\alpha'' \times \beta) = j^*(\delta(\alpha'') \times \beta).$$

**Proof.**
Let $X \to k$ be a $G$-projective resolution and $Y \to k$ an $H$-projective resolution. Then, since $X \otimes Y$ is $k(G \times H)$-projective,
$$0 \to \text{Hom}_{G \times H}(X \otimes Y, C) \to \text{Hom}_{G \times H}(X \otimes Y, A \otimes B)$$
$$\to \text{Hom}_{G \times H}(X \otimes Y, A \otimes B'') \to 0$$
is exact. Let $f \in \text{Hom}_G(X, A)$ be a *cocycle* representing $\alpha$ and let $g'' \in \text{Hom}_H(Y, B'')$ be a *cocycle* representing $\beta''$. Choose $g \in \text{Hom}_H(Y, B)$ mapping onto $g''$; then by naturality $f \times g$ maps onto $f \times g''$. Because the map $\times$ is a map of complexes, the rule for the differential in a tensor product yields
$$d(f \times g) = df \times g + (-1)^r f \times dg = (-1)^r f \times dg.$$
Since $dg'' = 0$, it follows from the exactness of
$$0 \to \text{Hom}_H(Y, B') \to \text{Hom}_H(Y, B) \to \text{Hom}_H(Y, B'') \to 0$$
that $dg$ is the image of a $g' \in \text{Hom}_H(Y, B')$ which represents $\delta\beta''$. Hence, $(-1)^r f \times g' \in \text{Hom}_{G \times H}(X \otimes Y, A \otimes B')$ maps onto $d(f \times g)$. From the commutativity of

$$\begin{array}{ccc}
& \text{Hom}_{G \times H}(X \otimes Y, A \otimes B') & \\
& j^* \downarrow & \searrow \\
0 \to \text{Hom}_{G \times H}(X \otimes Y, C) & \to \text{Hom}_{G \times H}(X \otimes Y, A \otimes B) &
\end{array}$$

it follows that $j^*((-1)^r f \times g')$ represents $\delta(\alpha \times \beta'')$. □

**Proposition 3.3.2.** *Let $G$ be a group, let*
$$0 \to A' \to A \to A'' \to 0$$
$$0 \to C' \to C \to C'' \to 0$$
*be short exact sequences of $G$-modules, and let $B$ be a $G$-module. Suppose there is a commutative diagram*

$$\begin{array}{ccccccc}
& A' \otimes B & \to & A \otimes B & \to & A'' \otimes B & \to 0 \\
& \downarrow & & \downarrow & & \downarrow & \\
0 \to & C' & \to & C & \to & C'' & \to 0
\end{array}$$

Then, using the cup products induced by the vertical maps, we have

$$\delta(\alpha''\beta) = \delta(\alpha'')\beta$$

for $\alpha'' \in H^r(G, A'')$ and $\beta \in H^s(G, B)$. Similarly, replacing the sequence of $A$'s by a single module and $B$ by a sequence, and supposing the appropriate diagram, we have

$$\delta(\alpha\beta'') = (-1)^r \alpha \delta(\beta'').$$

The most important connecting homomorphisms are those arising from the two exact sequences

$$0 \to \mathbf{Z} \xrightarrow{p} \mathbf{Z} \to \mathbf{Z}/p\mathbf{Z} \to 0$$

and

$$0 \to \mathbf{Z}/p\mathbf{Z} \to \mathbf{Z}/p^2\mathbf{Z} \to \mathbf{Z}/p\mathbf{Z} \to 0$$

where $p$ is a prime. We shall ordinarily denote the first one

$$\beta : H^n(G, \mathbf{Z}/p\mathbf{Z}) \to H^{n+1}(G, \mathbf{Z})$$

and the second one

$$\delta : H^n(G, \mathbf{Z}/p\mathbf{Z}) \to H^{n+1}(G, \mathbf{Z}/p\mathbf{Z}).$$

Both these maps are conventionally called Bockstein homomorphisms. $\delta$ is obtained by composing $\beta$ with the map induced by the projection $\mathbf{Z} \to \mathbf{Z}/p\mathbf{Z}$.

**Exercise 3.3.1.** Show $\delta$ is a derivation of the ring $H^*(G, \mathbf{Z}/p\mathbf{Z})$ and $\delta^2 = 0$.

The Bockstein provides a convenient way to reduce assertions about the cohomology of $\mathbf{Z}G$-modules to that of $(\mathbf{Z}/p\mathbf{Z} = \mathbf{F}_p)G$-modules.

**Example.** Suppose $P$ is cyclic of order $p$. Then

$$\begin{aligned} H^n(P, \mathbf{Z}) &= \mathbf{Z} & &\text{for } n = 0 \\ &= \mathbf{Z}/p\mathbf{Z} & &\text{for } n > 0 \text{ and even} \\ &= 0 & &\text{for } n \text{ odd} \\ H^n(P, \mathbf{Z}/p\mathbf{Z}) &= \mathbf{Z}/p\mathbf{Z} & &\text{for all } n. \end{aligned}$$

It follows that the short exact sequence

$$0 \to \mathbf{Z} \to \mathbf{Z} \to \mathbf{Z}/p\mathbf{Z} \to 0$$

yields a long exact sequence which starts off

$$0 \to \mathbf{Z} \xrightarrow{p} \mathbf{Z} \to \mathbf{Z}/p\mathbf{Z} \xrightarrow{0} H^1(P,\mathbf{Z}) = 0 \xrightarrow{p=0} H^1(P,\mathbf{Z}) = 0$$
$$\to H^1(P,\mathbf{Z}/p\mathbf{Z}) \xrightarrow{\beta} H^2(P,\mathbf{Z}) \xrightarrow{p=0} H^2(P,\mathbf{Z}) \to H^2(P,\mathbf{Z}/p\mathbf{Z}) \xrightarrow{\beta} \ldots$$

Hence, $\beta^1 : H^1(P, \mathbf{Z}/p\mathbf{Z}) \to H^2(P, \mathbf{Z})$ is an isomorphism. Moreover, composing with $H^2(P, \mathbf{Z}) \to H^2(P, \mathbf{Z}/p\mathbf{Z})$ shows that $\delta^1 : H^1(P, \mathbf{Z}/p\mathbf{Z}) \to H^2(P, \mathbf{Z}/p\mathbf{Z})$ is also an isomorphism. As in Section 3.2, a generator $\eta \in H^1(P, \mathbf{F}_p) = \text{Hom}(P, \mathbf{F}_p)$ may be characterized by $\eta(g) = 1$ where $P = \langle g \rangle$, and then $\xi = \delta\eta$ generates $H^2(P, \mathbf{F}_p)$.

If we put this information together with the product calculations in Section 3.2, we see that for $P$ cyclic of prime order $p$, $H^*(P, \mathbf{F}_p) \cong \mathbf{F}_p[\eta, \xi = \delta\eta \,|\, \eta^2 = 0]$ for $p$ odd and $H^*(P, \mathbf{F}_p) \cong \mathbf{F}_p[\eta]$ for $p = 2$. (Indeed, the corresponding result holds for any extension $k$ of $\mathbf{F}_p$.) Similarly, $H^*(P, \mathbf{Z}) = \mathbf{Z}[\xi \,|\, p\xi = 0]$ where $\xi = \beta\eta$, i.e, for positive degrees, $H^*(P, \mathbf{Z})$ is a polynomial ring over $\mathbf{F}_p = \mathbf{Z}/p\mathbf{Z}$ in the degree 2 generator $\xi$. (Of course, this fails in degree 0 where we have $\mathbf{Z}$.)

Another useful connecting homomorphism arises from the short exact sequence

$$0 \to \mathbf{Z} \to \mathbf{Q} \to \mathbf{Q}/\mathbf{Z} \to 0.$$

For $G$ finite cyclic, we may check by direct calculation that $H^n(G, \mathbf{Q}) = 0$ for $n > 0$. (See also Exercise 2.1.1.) We shall see later (Corollary 4.2.3) that, if $G$ is any finite group, then for any $G$-module $M$, $|G|H^n(G, M) = 0$, $n > 0$. Since $H^n(G, \mathbf{Q})$ is a rational vector space in any case, it vanishes for $n > 0$. It then follows that the connecting homomorphisms

$$H^n(G, \mathbf{Q}/\mathbf{Z}) \to H^{n+1}(G, \mathbf{Z})$$

are isomorphisms for $n > 0$. In particular, for a finite group

$$\text{Hom}(G, \mathbf{Q}/\mathbf{Z}) \cong H^1(G, \mathbf{Q}/\mathbf{Z}) \cong H^2(G, \mathbf{Z}).$$

## 3.4 The Universal Coefficient Theorem

Let $k$ be an appropriate ring, e.g. a field or a Dedekind domain, let $X$ be a $k$-free complex, and let $M$ be any $k$-module. Then the *Universal Coefficient Theorem* provides two exact sequences:

$$0 \to H_*(X) \otimes M \to H_*(X \otimes M) \to \text{Tor}_1^k(H_{*-1}(X), M) \to 0$$

and

$$0 \to \mathrm{Ext}^1_k(H_{*-1}(X), M) \to H^*(\mathrm{Hom}(X, M)) \to \mathrm{Hom}(H_*(X), M) \to 0.$$

(See Hilton and Stammbach (1971, Sections V.2–3) or Mac Lane (1963, Sections III.2, V.11) for details.) These sequences have many interesting applications to group cohomology, but we shall restrict attention here to the case $k$ is a field, in which the Tor and Ext terms vanish. Let $G$ be a group, $P \to k$ a $kG$-projective resolution, and $M$ a $kG$-module on which $G$ acts trivially. Then we have

$$P \otimes_{kG} k \otimes M \cong P \otimes_{kG} M$$
$$\mathrm{Hom}(P \otimes_{kG} k, M) \cong \mathrm{Hom}_{kG}(P, M)$$
$$\mathrm{Hom}_{kG}(P, k) \otimes M \cong \mathrm{Hom}_{kG}(P, M).$$

The last isomorphism requires some mild finiteness assumptions on $P$ and is defined by $f \otimes m \mapsto F$ where $F(p) = f(p)m$.

With $X = P \otimes_{kG} k$, the Universal Coefficient Theorem provides isomorphisms

$$H_*(G, M) \cong H_*(G, k) \otimes M$$
$$H^*(G, M) \cong \mathrm{Hom}(H_*(G, k), M).$$

With $X = \mathrm{Hom}_{kG}(P, k)$, which is a *cochain complex*, we obtain

$$H^*(G, k) \otimes M \cong H^*(G, M).$$

This last isomorphism is quite easy to describe. For $\alpha \in H^*(G, k)$, let $f : P \to k$ be a representative cocycle. Then, if $f \otimes m \mapsto F$ as above, it is easy to check that $F \in \mathrm{Hom}_{kG}(P, M)$ is also a cocycle. Map $\alpha \otimes m$ to the cohomology class represented by $F$.

One case which will interest us below is the following. If $k$ is a field of characteristic $p > 0$, then we may view it as an algebra over $\mathbf{F}_p$. Hence, we obtain an isomorphism

$$H^*(G, \mathbf{F}_p) \otimes_{\mathbf{F}_p} k \cong H^*(G, k),$$

and it is easy to see that this is an isomorphism of $k$-algebras. (Of course, we could do the same for any field extension.) Let $\delta : H^*(G, \mathbf{F}_p) \to H^{*+1}(G, \mathbf{F}_p)$ be the Bockstein homomorphism defined in Section 3.3. This induces a $k$-linear map $\delta \otimes \mathrm{Id} : H^*(G, k) \to H^{*+1}(G, k)$ which we shall also call the Bockstein and usually just denote by $\delta$.

There is another way to extend the Bockstein to a map $\Delta : H^*(G, k) \to H^{*+1}(G, k)$. To this end, construct a short exact sequence of abelian groups

$$0 \to k \to W_2(k) \to k \to 0$$

as follows. Let $W_2(k) = k \times k$ as a set. Define a group operation on $W_2(k)$ by
$$(x_0, x_1) + (y_0, y_1) = (x_0 + y_0, x_1 + y_1 - \Phi(x_0, y_0))$$
where
$$\Phi(X, Y) = \sum_{i=1}^{p-1} \frac{1}{p}\binom{p}{i} X^i Y^{p-i}$$
is a homogeneous polynomial of degree $p$ over the field $\mathbf{F}_p$. Note that formally
$$\Phi(X, Y) = \frac{(X+Y)^p - X^p - Y^p}{p}.$$
(It is also possible to give $W_2(k)$ a ring structure so that $W_2(k) \to k$ is a ring homomorphism. It is called the ring of truncated *Witt vectors* of length 2.)

Define a map $k \to W_2(k)$ by $x \mapsto (0, x)$ and $W_2(k) \to k$ by $(x_0, x_1) \mapsto x_0$. Some tedious calculations show that $W_2(k)$ is an abelian group under the given operation, and the given maps fit it into a short exact sequence. We define $\Delta : H^*(G, k) \to H^{*+1}(G, k)$ to be the resulting connecting homomorphism.

**Exercise 3.4.1.** Show that the restriction of $H^*(G, k) \xrightarrow{\Delta} H^{*+1}(G, k)$ to $H^*(G, \mathbf{F}_p) \cong H^*(G, \mathbf{F}_p) \otimes_{\mathbf{F}_p} 1$ agrees with the Bockstein defined for $\mathbf{F}_p$. Hint: Compare the short exact sequences

$$\begin{array}{ccccccccc}
0 & \to & \mathbf{F}_p & \to & W_2(\mathbf{F}_p) & \to & \mathbf{F}_p & \to & 0 \\
& & \downarrow & & \downarrow & & \downarrow & & \\
0 & \to & k & \to & W_2(k) & \to & k & \to & 0
\end{array}$$

and show that $W_2(\mathbf{F}_p) \cong \mathbf{Z}/p^2\mathbf{Z}$.

Note that $\Delta$ is not generally a $k$-homomorphism, but is instead semilinear with respect to the Frobenius endomorphism $\phi : k \to k$ defined by $\phi(a) = a^p$, i.e.
$$\Delta(a\alpha) = a^p \Delta(\alpha) \qquad \text{for } a \in k, \alpha \in H^*(G, k).$$

However, if the field $k$ is *perfect*, then $\phi$ is an automorphism, and if we define
$$\delta' = (\phi^*)^{-1} \circ \Delta$$
where $\phi^* : H^*(G, k) \to H^*(G, k)$ is induced by $\phi$, then $\delta'$ is $k$-linear. Clearly, $\delta'$ also restricts to the ordinary Bockstein on $H^*(G, \mathbf{F}_p)$.

**Exercise 3.4.2.** Show that if $k$ is perfect, then $\delta' = \delta \otimes \text{Id}$.

## 3.5 Cohomology rings of direct products and abelian groups

Let $G = H \times K$. Then $H^*(H,k) \otimes H^*(K,k)$ becomes a $k$-algebra if we define

$$(\alpha \otimes \beta)(\gamma \otimes \delta) = (-1)^{\deg \beta \, \deg \gamma} \alpha\gamma \otimes \beta\delta$$

for homogeneous elements and then extend by linearity. Moreover, it is not hard to see that the external product map

$$\times : H^*(H,k) \otimes H^*(K,k) \to H^*(H \times K, k)$$

is an algebra homomorphism. (Choose resolutions and diagonal chain maps $D_X : X \to X \otimes X$ and $D_Y : Y \to Y \otimes Y$ for $H$ and $K$ respectively, and then use $D_X \otimes D_Y$ followed by twisting for a diagonal map $X \otimes Y \to (X \otimes Y) \otimes (X \otimes Y)$. The twisting introduces the sign.) In particular, if $k$ is chosen properly, we know from our discussion of the Künneth Theorem in Section 2.5 that this algebra homomorphism is a monomorphism. In particular, if $k$ is a field, $\times$ provides an algebra isomorphism

$$H^*(H,k) \otimes H^*(K,k) \cong H^*(H \times K, k).$$

Combining the above remarks with our discussion of cohomology rings of cyclic groups allows us in principle to calculate the cohomology ring of any finite abelian group.

For example, let $G = P^d$ where $P$ is cyclic of prime order $p$ so that $G$ is *elementary abelian*, and let $k$ be a field of characteristic $p$. Then

$$H^*(G,k) \cong H^*(P,k)^{\otimes d}$$

where the tensor algebra on the right involves signs, as above, for elements of odd degree. In particular, if $p$ is odd, we obtain

$$H^*(G,k) \cong k[\eta_1, \xi_1] \otimes \cdots \otimes k[\eta_d, \xi_d]$$
$$\cong k[\eta_1, \ldots, \eta_d, \xi_1, \ldots, \xi_d]$$

where $\deg \eta_i = 1, \deg \xi_i = 2$, and $\eta_i^2 = 0$. This may also be rewritten

$$H^*(G,k) \cong k[\eta_1, \ldots, \eta_d] \otimes k[\xi_1, \ldots, \xi_d]$$

where the first term on the right is an exterior algebra over $k$ and the second term is a polynomial algebra. As in the cyclic case, we may take $\{\eta_i \mid i = 1, \ldots, d\}$ to be a basis for $\mathrm{Hom}(G,k) = H^1(G,k)$ and $\xi_i = \delta\eta_i$. (Just use the fact that $\delta$ commutes with homomorphisms induced by group

homomorphisms and project onto the $i$th factor.) Similarly, for $p = 2$, we have more simply
$$H^*(G, k) \cong k[\eta_1, \ldots, \eta_d],$$
a polynomial algebra generated by elements of degree 1. These calculations work just as well for arbitrary finite abelian groups except for the remarks about the Bocksteins.

If $k$ is not a field, the situation is much more involved. For $k = \mathbf{Z}$, a complete description of this cohomology *ring* $H^*(G, \mathbf{Z})$ was first worked out by Chapman (1982) for abelian groups of odd order and by Townsley Kulich (1988) for abelian groups of arbitrary order. For $G = P^d$ we still have a ring monomorphism
$$H^*(P, \mathbf{Z})^{\otimes d} \to H^*(P^d, \mathbf{Z}).$$

As in Section 3.3, $H^*(P, \mathbf{Z}) = \mathbf{Z}[\xi \mid p\xi = 0]$ (where $\xi = \beta\eta$), so $H^*(G, \mathbf{Z})$ contains a subring $\mathbf{Z}[\xi_1, \ldots, \xi_d \mid p\xi_i = 0]$, where $\xi_i = \beta(\eta_i), i = 1, \ldots, d$, form a basis for $H^2(G, \mathbf{Z}) \cong \mathrm{Hom}_{\mathbf{Z}}(G, \mathbf{F}_p) \cong \mathrm{Hom}(G, \mathbf{Q}/\mathbf{Z})$. This subring may also be described as the symmetric algebra $S_{\mathbf{Z}}(\hat{G})$ (where $\hat{G} = \mathrm{Hom}(G, \mathbf{Q}/\mathbf{Z}))$ and in positive degrees it behaves like a polynomial ring over $\mathbf{F}_p$ in $d$ indeterminates $\xi_1, \xi_2, \ldots, \xi_d$.

# 4

# Relations to cohomology of subgroups

## 4.1 Restriction and the Eckmann–Shapiro Lemma

Let $G$ be a group and $H$ a subgroup. If $A$ is a $kG$-module, it is also a $kH$-module by restriction, and by functorality there is an induced map $H^*(G, A) \to H^*(H, A)$ which is usually called *restriction* and denoted $\mathrm{res}_{G \to H}$. If $X \to k$ is a $kG$-projective resolution, since $kG$ is $kH$-free, $X \to k$ is also a $kH$-projective resolution. Hence, restriction is induced from the inclusion

$$\mathrm{Hom}_G(X, A) \to \mathrm{Hom}_H(X, A).$$

This observation simplifies many proofs which were originally done in a very cumbersome manner using the bar resolution.

If $g \in G$, there is a map

$$g^* : \mathrm{Hom}_H(X, A) \to \mathrm{Hom}_{gHg^{-1}}(X, A)$$

defined by $g^*(f)(x) = gf(g^{-1}x)$. (The reader should check that

$$g^*(f)(ghg^{-1}x) = ghg^{-1}g^*(f)(x)$$

for $h \in H$ and $x \in X$.) This induces a morphism

$$g^* : H^*(H, A) \to H^*(gHg^{-1}, A).$$

(In fact, $g^* = (\phi, f)^*$ where $\phi : gHg^{-1} \to H$ is defined by $\phi(h') = g^{-1}h'g$ and $f : A \to A$ is defined by $f(a) = ga$. See Section 1.1.) It is easy to check that $(g_1 g_2)^* = g_1^* g_2^*$, so in particular if $H$ is normal in $G$, an action of $G$ on $H^*(H, A)$ is induced. It is clear from the definitions that in the normal case $h^* = \mathrm{Id}$ for $h \in H$, i.e. $H$ acts trivially on $H^*(H, A)$ so that the latter is actually a $k(G/H)$-module. In particular, if $G = H$, we have the following proposition.

**Proposition 4.1.1.** *$G$ acts trivially on $H^*(G, A)$.*

Let $G$ be a group, $H$ a subgroup, and $A$ a $kH$-module. Consider

$$kG \otimes_{kH} A$$

where *in this case only* we treat $kG$ as a *right* $kH$-module by multiplying on the right by elements of $H$ and we use the usual tensor product of a right module and a left module. Let $G$ act on $kG \otimes_{kH} A$ by $g(x \otimes a) = gx \otimes a$. The resulting $kG$-module is called the *induced module*. Similarly, we define the *coinduced module* to be

$$\mathrm{Hom}_{kH}(kG, A)$$

where here $H$ acts on $kG$ on the left as usual, and we let $G$ act on $\mathrm{Hom}_{kH}(kG, A)$ by $(gf)(x) = f(xg)$. Note that in either case we need to have $G$ and $H$ act on *opposite* sides of $kG$ so that the two actions will *commute*. Switching from right modules to left modules by means of the inverse map will not deal with that problem.

**Proposition 4.1.2.** *If $(G : H) < \infty$ then there is a natural isomorphism of $kG$-modules*

$$kG \otimes_{kH} A \cong \mathrm{Hom}_{kH}(kG, A)$$

We shall sometimes use the notation $A\uparrow_H^G$ to denote either module in the case of finite index.

**Proof.** Let $T$ be a set of left coset representatives of $H$ in $G$. Define a map $\mathrm{Hom}_{kH}(kG, A) \to kG \otimes_{kH} A$ by $f \mapsto \sum_{t \in T} t \otimes f(t^{-1})$. Similarly, define a map $kG \otimes_{kH} A \to \mathrm{Hom}_{kH}(kG, A)$ by $\sum_{t \in T} t \otimes a_t \mapsto f$ where $f(ht^{-1}) = ha_t$. (Note that each element of $kG \otimes_{kH} A$ is uniquely representable in the given form, and $\{t^{-1} \mid t \in T\}$ is a set of right coset representatives.) It is easy to check that these are $G$-homomorphisms and they are clearly inverses. □

For $A$ a $kH$-module, define

$$j : \mathrm{Hom}_{kH}(kG, A) \to A = \mathrm{Hom}_{kH}(kH, A)$$

by $j(f) = f(1)$. (In effect, $j$ just restricts $f$ to $kH$.) It is easy to check that $j$ is a $kH$-homomorphism. Let $\iota : H \to G$ denote the inclusion.

**Proposition 4.1.3** (Eckmann–Shapiro). *For $H$ a subgroup of $G$ and $A$ a $kH$-module, with the notation as above,*

$$j^* \,\mathrm{res}_{G \to H} = (\iota, j)^* : H^*(G, \mathrm{Hom}_H(kG, A)) \to H^*(H, A)$$

is an isomorphism.

**Proof.** $(\iota, j)$ induces

$$\mathrm{Hom}_G(X, \mathrm{Hom}_H(kG, A)) \to \mathrm{Hom}_H(X, \mathrm{Hom}_H(kH, A)) \to \mathrm{Hom}_H(X, A)$$

which comes down to $F \mapsto f$, where $f(x) = F(x)(1)$, and it is clear what inverse to use. □

The lemma was first stated by Eckmann (1953), but it seems to have been discovered independently by others at about the same time. Hearsay has it that Weil attributed the result to Arnold Shapiro, who was embarrassed to have his name attached to what seemed such a minor result. Given its wide use, he may have thought better of it later. The result, like Schur's Lemma and similar results, seems to have an importance which belies its seeming superficiality.

There are several other Eckmann–Shapiro type lemmas. First, the homology version asserts that $H_*(G, kG \otimes_{kH} A) \cong H_*(H, A)$ for every $H$-module A. Moreover, one can do the transition using the first variable in Ext to conclude that

$$\mathrm{Ext}_{kG}(kG \otimes_{kH} M, N) \cong \mathrm{Ext}_{kH}(M, N)$$

for every $kH$-module $M$ and $kG$-module $N$. In particular,

$$H^*(H, A) \cong \mathrm{Ext}^*_{kG}(kG \otimes_{kH} k, A)$$

for every $kG$-module $A$. (Similar remarks apply for homology and Tor.)

**Exercise 4.1.1.** Let $G$ be a group and $H$ a subgroup. Define

$$\phi : gHg^{-1} \to H \quad \text{by} \quad \phi(h') = g^{-1}h'g$$
$$f : A \to A \quad \text{by} \quad f(a) = ga.$$

(a) Show that $f(\phi(h')a) = h'\phi(a)$.
(b) Show that $g^* = (\phi, f)^* : H^*(H, A) \to H^*(gHg^{-1}, A)$.
Hint: Choose a $kG$-projective resolution $X \to k$ so that it is also a $kH$ and $k(gHg^{-1})$-projective resolution. Define $\Phi : X \to X$ by $\Phi(x) = g^{-1}x$ and show that $\Phi(h'x) = \phi(h')\Phi(x)$.

**Exercise 4.1.2.** Let $H$ and $K$ be groups which share a common subgroup $A$. A *free product with amalgamated subgroup* is a group $G = K *_A H$ containing isomorphic copies of $H$ and $K$ in such a way that the diagram

$$\begin{array}{ccc} A & \longrightarrow & H \\ \downarrow & & \downarrow \\ K & \longrightarrow & G \end{array}$$

presents $G$ as a pushout. We have obvious maps

$$\begin{array}{ccc} \mathbf{Z}(G/A) & \xrightarrow{\iota'} & \mathbf{Z}(G/H) \\ {\scriptstyle \iota''}\downarrow & & \downarrow{\scriptstyle \epsilon'} \\ \mathbf{Z}(G/K) & \xrightarrow{\epsilon''} & \mathbf{Z} \end{array}$$

Assume (Swan 1969) that the sequence

$$0 \to \mathbf{Z}(G/A) \xrightarrow{\iota} \mathbf{Z}(G/H) \oplus \mathbf{Z}(G/K) \xrightarrow{\kappa} \mathbf{Z} \to 0$$

is exact, where $\iota(x) = (\iota'(x), -\iota''(x))$ and $\kappa(x', x'') = \epsilon'(x') + \epsilon''(x'')$. Using an appropriate Eckmann–Shapiro Lemma, derive a Mayer–Vietoris sequence

$$0 \to M^G \to M^K \oplus M^H \to M^A \to$$
$$H^1(G, M) \to H^1(H, M) \oplus H^1(K, M) \to H^1(A, M) \to \cdots$$

and similarly for homology.

**Exercise 4.1.3.** Show that the short exact sequence in the previous exercise is exact. Here are some hints from Bieri (1976, Section 2). Choose generators $\{x_p, t_q\}$ for $H$ where $\{t_q\}$ generates $A$ and no $x_p \in A$. Similarly, choose generators $\{y_r, t_q\}$ for $K$. Then $\{x_p, t_q, y_r\}$ generates $G$. To show $\iota$ is a monomorphism, suppose $\iota(x) = 0$, and in $u = (\sum_{w \in G} n_w w)\bar{1}$ choose a $w$ of maximal length when expressed minimally in terms of the generators. Note that the last generator may be assumed to be either an $x_p$ or a $y_r$. To show Ker $\kappa$ = Im $\iota$, argue as follows. Note that any element of $\mathbf{Z}(G/H)$ can be expressed as $u = (\sum n_w w)\bar{1} = (\sum n_w(w-1))\bar{1} + \epsilon'(u)\bar{1}$. Use the fact that $w \mapsto w - 1$ is a derivation to express the first part as a linear combination of the $y_i - 1$ with coefficients in $\mathbf{Z}G$. Repeat for $\mathbf{Z}(G/K)$, and consider an appropriate sum for $\mathbf{Z}(G/A)$.

## 4.2 Transfer or corestriction

Let $G$ be a group, $H$ a subgroup of finite index, and $M$ a $kG$-module. Let $S$ be a set of left coset representatives of $H$ in $G$. If $X \to k$ is a $kG$-projective resolution, then it is also a $kH$-projective resolution. Define $T_{G/H} : \mathrm{Hom}_H(X, M) \to \mathrm{Hom}_G(X, M)$ by

$$T_{G/H}(f) = \sum_{s \in S} sf.$$

Note that the result does not depend on the choice of coset representatives. For, if $h \in H$, we have

$$(shf)(x) = shf(h^{-1}s^{-1}x) = sf(s^{-1}x)$$

since $f \in \mathrm{Hom}_H(X, M)$. It is not hard to see that $T_{G/H}$ commutes with the differentials in the two complexes so it induces a homomorphism

$$\mathrm{cor}_{H \to G} : H^*(H, M) \to H^*(G, M)$$

which is called *corestriction* or *transfer*. It is not hard to see that this morphism does not depend on the $kG$-projective resolution. For homology, there is correspondingly a morphism $H_*(G, M) \to H_*(H, M)$ induced by

$$x \otimes m \mapsto \sum_{s \in S'} sx \otimes sm$$

where $S'$ is a set of *right* coset representatives. This map should probably just be called *transfer*.

**Proposition 4.2.1.** *If $H$ is a subgroup of $K$ and $K$ is a subgroup of $G$ with $H$ of finite index in $G$, then*

$$\mathrm{cor}_{H \to G} = \mathrm{cor}_{K \to G}\, \mathrm{cor}_{H \to K}.$$

**Proof.** Let $S$ be a set of left coset representatives of $H$ in $K$ and $R$ a set of left coset representatives of $K$ in $G$. Then the set of products $RS$ is a set of left coset representatives of $H$ in $G$. Calculating the corestriction with this set of coset representatives gives the result. □

**Proposition 4.2.2.** *If $H$ is a subgroup of $G$ of finite index and $M$ is a $kG$-module, then*

$$\mathrm{cor}_{H \to G}\, \mathrm{res}_{G \to H} = (G : H)\, \mathrm{Id}.$$

**Proof.** $T_{G/H}(f) = \sum_{s \in S} sf = \sum_{s \in S} f = (G:H)f$. □

**Corollary 4.2.3.** *If $G$ is a finite group and $M$ is a $kG$-module, then for $n > 0$,*

$$|G|\, H^n(G, M) = \{0\}.$$

*If in addition $M$ is finitely generated as a $\mathbf{Z}$-module, then $H^n(G, M)$ is finite for $n > 0$.*

**Proof.** Apply the above formula with $H = 1$, and use $H^n(1, M) = 0$ for $n > 0$. □

**Proposition 4.2.4.** *Let $H$ be a subgroup of $G$ of finite index and let $A$ and $B$ be $kG$-modules. Then for $\alpha \in H^*(G, A)$ and $\beta \in H^*(H, B)$,*

$$\mathrm{cor}_{H \to G}(\mathrm{res}_{G \to H}(\alpha)\beta) = \alpha \, \mathrm{cor}_{H \to G}(\beta).$$

**Proof.** Let $X \to k$ be a $G$-projective resolution. Let $f \in \mathrm{Hom}_G(X, A)$ represent $\alpha$ and $g \in \mathrm{Hom}_H(X, B)$ represent $\beta$. Then $f \in \mathrm{Hom}_H(X, A)$ where it represents $\mathrm{res}\,\alpha$. Through the diagonal action, we may view $X \otimes X \to k$ as either a $G$-projective resolution or as an $H$-projective resolution. Then the identity map $X \otimes X \to X \otimes X$ may be used to compute products, where on the left the complex is viewed as a $G$ (similarly $H$) complex and on the right as a $G \times G$ (similarly $H \times H$) complex. With these assumptions, $f \times g \in \mathrm{Hom}_H(X \otimes X, A \otimes B)$ represents $(\mathrm{res}\,\alpha)\beta \in H^*(H, A \otimes B)$. Since $\mathrm{cor}((\mathrm{res}\,\alpha))\beta$ is represented by $T_{G/H}(f \times g) = \sum_s s(f \times g)$, and since $sf = f$, it suffices to prove

$$s(f \times g) = sf \times sg \quad \text{for } s \in G, f \in \mathrm{Hom}_H(X, A), g \in \mathrm{Hom}_H(X, B).$$

This follows by direct calculation if we keep straight the assumptions about group actions and other definitions:

$$\begin{aligned}
s(f \times g)(x \otimes y) &= s((f \times g)(s^{-1}x \otimes s^{-1}y)) \\
&= s(f(s^{-1}x) \otimes g(s^{-1}y)) = sf(s^{-1}x) \otimes sg(s^{-1}y) \\
&= (sf)(x) \otimes (sg)(y) = ((sf) \times (sg))(x \otimes y).
\end{aligned}$$

Note that the above calculation may be extended to show that the rule $s(fg) = (sf)(sg)$ holds even if we carry the product back by a diagonal map to the resolution $X$. □

The most important application of corestriction for finite groups is that it allows us to relate the cohomology of $G$ to that of $H$ in the case $(G:H)$ and $|H|$ are relatively prime. (In that case $H$ is called a *Hall subgroup* of $G$. The most interesting Hall subgroups are the Sylow subgroups for various primes.) Denote by $H^n(G, M; m)$ the $m$-primary component of $H^n(G, M)$. If $G$ is finite, and $|G| = mr$ where $(m, r) = 1$, then since $H^n(G, M)$ is an $mr$ torsion group for $n > 0$, it is the direct sum of its $m$-primary and $r$-primary components.

**Proposition 4.2.5.** *If $H$ is a Hall subgroup of $G$ and $M$ is a $G$-module, then, for $n > 0$,*

$$H^n(H, M) = \mathrm{Ker}\,\mathrm{cor}_{H \to G} \oplus \mathrm{Im}\,\mathrm{res}_{G \to H}$$

*where* $\mathrm{Im}\,\mathrm{res}_{G \to H} \cong H^n(G, M; |H|) = \mathrm{Im}\,\mathrm{cor}_{H \to G}.$

**Proof.** Since $H^n(H, M)$ is $|H|$-torsion, it follows that

$$\mathrm{Im}\,\mathrm{cor}_{H \to G} \subseteq H^n(G, M; |H|).$$

Hence, we may restrict the formula

$$\text{cor}_{H\to G}\,\text{res}_{G\to H} = (G:H)\,\text{Id}$$

to $H^n(G, M; |H|)$ and on that group the right hand side is an isomorphism since $(G:H)$ is relatively prime to $|H|$. This gives the desired splitting. □

We can go further in characterizing $\text{Im}\,\text{res}_{G\to H} \cong H^n(G, M; |H|)$ in the case of a Hall subgroup. For this we need a formula which tells us what happens if we corestrict from a subgroup $H$ and then restrict back to another (possibly the same) subgroup $K$. In that situation, the set of left cosets $G/H$ may be viewed as a $K$ set and as such decomposes into disjoint orbits $K\bar{x}$ where $x$ ranges over an appropriate subset $D$ of representatives in $G$. $D$ is called a set of *double coset representatives*, and we have a decomposition $G = \cup_{x\in D} KxH$ into disjoint subsets $KxH$ called *double cosets*. The stabilizer of $\bar{x}$ in $K$ is $\{k \in K \mid kxH = xH\} = K \cap xHx^{-1}$. Hence, $K\bar{x} \cong K/K \cap xHx^{-1}$ is a $K$-set. (Let $k\bar{x} \leftrightarrow k(K \cap xHx^{-1})$.) Let $S(x)$ be a complete set of left coset representatives of $K \cap xHx^{-1}$ in $K$, and let $S = \cup_{x\in D} S(x)x$ so that $S$ is a complete set of coset representatives of $H$ in $G$.

Let $X \to k$ be a $kG$-projective resolution and let $f \in \text{Hom}_H(X, M)$ represent $\alpha \in H^*(H, M)$. Then

$$\sum_{s\in S} sf = \sum_{x\in D} \left( \sum_{t\in S(x)} txf \right).$$

If we restrict now to $K$, we see that $\sum_{t\in S(x)} txf$ represents

$$\text{cor}_{K\cap xHx^{-1}\to K}(x^*\alpha).$$

We have now proved the following formula.

**Theorem 4.2.6.** *Let $G$ be a group, $H$ a subgroup of finite index, and $K$ another subgroup. Let $D$ be a set of double coset representatives (so that $G = \cup_{x\in D} KxH$ is a disjoint union.) Then for $\alpha \in H^*(H, M)$, we have*

$$\text{res}_{G\to K}(\text{cor}_{H\to G}(\alpha)) = \sum_{x\in D} \text{cor}_{K\cap xHx^{-1}\to K}(\text{res}_{xHx^{-1}\to K\cap xHx^{-1}}(x^*\alpha))$$

$$= \sum_{x\in D} \text{cor}_{K\cap xHx^{-1}\to K}(x^*(\text{res}_{H\to H\cap x^{-1}Kx}(\alpha))).$$

**Corollary 4.2.7.** *Let $H$ be a Hall subgroup of $G$. Then $\alpha \in H^n(H, M)$ is in $\text{Im}\,\text{res}_{G\to H}$ if and only if*

$$\text{res}_{H\to H\cap xHx^{-1}}(\alpha) = \text{res}_{xHx^{-1}\to H\cap xHx^{-1}}(x^*\alpha) \tag{4.1}$$

for every $x \in D$ where $G = \cup_{x \in D} HxH$ is a double coset decomposition.

Elements satisfying the condition (4.1) are called *stable*.

**Proof.** Suppose $\alpha = \text{res}_{G \to H}(\beta)$. Then

$$x^*(\alpha) = x^*(\text{res}_{G \to H}(\beta)) = \text{res}_{G \to xHx^{-1}}(x^*\beta) = \text{res}_{G \to xHx^{-1}}(\beta)$$

by naturality and since $x \in G$ acts trivially on $H^*(G, M)$. Restricting further to the common subgroup $H \cap xHx^{-1}$ gives the desired condition for any $x \in G$. Conversely, suppose $\alpha \in H^n(H, M)$ is stable. Then

$$\text{cor}_{H \cap xHx^{-1} \to H}(\text{res}_{xHx^{-1} \to H \cap xHx^{-1}}(x^*\alpha))$$
$$= \text{cor}_{H \cap xHx^{-1} \to H}(\text{res}_{H \to H \cap xHx^{-1}}(\alpha)) = (H : H \cap xHx^{-1})\alpha,$$

so using the double coset formula yields

$$\text{res}_{G \to H} \text{cor}_{H \to G}(\alpha) = \sum (H : H \cap xHx^{-1})\alpha = (G : H)\alpha.$$

Since $(G : H)$ is relatively prime to $|H|$, $m(G : H) \equiv 1 \mod |H|$ for an appropriate integer $m$, and it follows that $\alpha \in \text{Im} \, \text{res}_{G \to H}$. □

Note that the validity of (4.1) for all $x \in D$ is equivalent to its validity for all $x \in G$.

As mentioned previously, the major application of these results is to the case of a $p$-Sylow subgroup $P$ of a finite group $G$. Then, for $n > 0$, $H^n(G, M; p)$ is isomorphic (under restriction) to the $k$-submodule of $H^n(P, M)$ of stable elements. Unfortunately, it is usually quite difficult to check stability, so this result is not as useful as it might appear to be. An exception is the case $P$ is *abelian* and $G$ acts *trivially* on $M$.

**Theorem 4.2.8.** *Let $G$ be a finite group, $M$ a trivial $G$ module, and $p$ a prime. Suppose the $p$-Sylow subgroups of $G$ are abelian, and let $P$ be one such Sylow subgroup. Then the set of stable elements of $H^n(P, M)$ is $H^n(P, M)^N$ where $N = N_G(P)$ is the normalizer of $P$ in $G$.*

This is a special case of a theorem of Swan (1960b) which generalizes the '2nd Theorem of Grün' (Hall 1959, Theorem 14.4.5) in finite group theory. (Swan claimed originally that the theorem holds for arbitrary coefficient modules, but that is wrong and was corrected in Swan (1962, Section 3).)

**Proof.** It is clear that every stable element is contained in $H^n(P, M)^N$. Suppose conversely that $\alpha \in H^n(G, M)^N$. Let $Q = P \cap xPx^{-1}$, and consider $C = C_G(Q)$, the centralizer of $Q$ in $G$. Since both $P$ and $xPx^{-1}$ are $p$-Sylow subgroups of $C$, it follows that there is a $y \in C$ such that

$xPx^{-1} = yPy^{-1}$ whence $y^{-1}x \in N$. By assumption $\alpha$ is fixed by $(y^{-1}x)^*$ from which we conclude that $x^*(\alpha) = y^*(\alpha)$. However,

$$\operatorname{res}_{yPy^{-1} \to Q} y^*(\alpha) = y^*(\operatorname{res}_{P \to Q} \alpha) = \operatorname{res}_{P \to Q} \alpha$$

since $C$ acts trivially on $Q$. It follows that $\alpha$ is stable for $x$. □

**Exercise 4.2.1.** Let $|G|$ be invertible in $k$. Show that $H^n(G, A) = 0$ for $n > 0$ for all $k$-modules $A$. In particular, show that if $G$ is finite and $k$ is a field of characteristic 0, then $H^n(G, k) = 0$ for $n > 0$.

**Exercise 4.2.2.** Let $G$ be a finite group and $A$ a finite abelian group such that $|G|$ and $|A|$ are relatively prime. Show that every group extension $A \to U \to G$ splits. (This is a special case of the *Schur–Zassenhaus Theorem* which asserts the same thing without the assumption that $A$ is abelian. The general case follows fairly easily from the abelian case. See Mac Lane (1963, Chapter IV, Theorem 10.5).)

**Exercise 4.2.3.** Using the sequence $0 \to \mathbf{Z} \to \mathbf{Q} \to \mathbf{Q}/\mathbf{Z} \to 0$, show that $H^2(G, \mathbf{Z}) \cong \operatorname{Hom}(G, \mathbf{Q}/\mathbf{Z}) \cong \operatorname{Hom}(G/G', \mathbf{Q}/\mathbf{Z})$.

**Exercise 4.2.4.** Let $H$ be a subgroup of $G$ of finite index, and let $0 \to A' \to A \to A'' \to 0$ be a short exact sequence of $G$-modules. Show that the connecting homomorphisms are consistent with corestriction, i.e.

$$\begin{array}{ccc} H^n(G, A'') & \xrightarrow{\delta} & H^{n+1}(G, A') \\ {\scriptstyle \operatorname{cor}} \uparrow & & \uparrow {\scriptstyle \operatorname{cor}} \\ H^n(H, A'') & \xrightarrow{\delta} & H^{n+1}(H, A') \end{array}$$

commutes.

**Exercise 4.2.5.** Let $p$ be a prime and let $G = S_p$, the symmetric group of degree $p$. Then each $p$-Sylow subgroup of $G$ is cyclic of order $p$. (One such is generated by the cycle $(1\,2\,\ldots\,p)$.) Hence, $H^*(P, \mathbf{Z}) \cong \mathbf{Z}[\eta]$ where $\deg \eta = 2$ and $p\eta = 0$. Show that $N_G(P)/P \cong (\mathbf{Z}/p\mathbf{Z})^*$ and it acts on $P \cong \mathbf{Z}/p\mathbf{Z}$ in the obvious way. Conclude that $H^*(S_p, \mathbf{Z}; p) \cong \mathbf{Z}[\eta^{p-1}]$.

# 5
# Cohomology of wreath products

Wreath products (defined below) seem to be ubiquitous in group theory and its applications. They appear implicitly in the work of 19th century group theorists, but were first defined explicitly by Polya when studying combinatorial problems associated with classifying organic molecules (see Polya and Read 1987, p.99). They were studied systematically in a series of papers by Kaloujnine and Krasner (1948, 1950, 1951a, 1951b). The Sylow subgroups of symmetric groups (Hall 1959, Section 5.9) are formed from wreath products, as are the Sylow subgroups of $Gl(n, \mathbf{F}_q)$ (and other linear groups) away from the characteristic of the field (Weir 1955). The cohomology of wreath products has played an important role in the development of the subject, in part because of interest in the cohomology of such groups, but also because wreath products are closely connected with the notion of transfer (corestriction) and its generalizations. We shall explore these notions in what follows.

## 5.1 Tensor induced modules

Let $G$ be a group, $H$ a subgroup of finite index, and $M$ a $kH$-module. We wish to define a multiplicative analogue of the induced module called the *tensor induced module* or *monomial module*. First we review the definition of the ordinary induced module. Let $T$ be a set of left coset representatives of $H$ in $G$. Then

$$M\uparrow_H^G = kG \otimes_{kH} M \cong \bigoplus_{t \in T} tkH \otimes_{kH} M \cong \bigoplus_{t \in T} t \otimes M.$$

We want to describe the action of an element $g \in G$ in terms of the right hand side. Let

$$gt = \overline{gt}\, h_{g,t}$$

where $h_{g,t} \in H$, and $\overline{s}$ denotes the coset representative in $T$ for $s \in G$. This action can be further described as follows. The choice of the set $T$ yields set theoretic bijections

$$G \cong T \times H \cong G/H \times H,$$

and left multiplication by an element $g \in G$ translates into a combination of the permutation $\pi(g) : t \mapsto \overline{gt}$ in $\mathcal{S}(T) \cong \mathcal{S}(G/H)$ with $|T|$ permutations in $\mathcal{S}(H)$ produced by left multiplication by the elements $h_{g,t}$, $t \in T$. With this notation, we have

$$g\left(\sum_{t \in T} t \otimes m_t\right) = \sum_{t \in T} \overline{g}th_{g,t} \otimes m_t = \sum_{t \in T} \pi(g)(t) \otimes h_{g,t}m_t$$

$$= \sum_{t \in T} t \otimes h_{g,\pi(g)^{-1}(t)} m_{\pi(g)^{-1}(t)}.$$

The above discussion shows how to identify $M\uparrow_H^G$ with $\bigoplus_{t \in T} M$ as a $G$-module. We can apply the same idea to the module

$$M^{\otimes T} = \bigotimes_{t \in T} M$$

formed by tensoring $(G:H)$ copies of $M$ over $k$. More precisely, let

$$g(\otimes_{t \in T} m_t) = \otimes_{t \in T} h_{g,\pi(g^{-1})t} m_{\pi(g^{-1})t}.$$

It is fairly clear from the above discussion that this defines an action although the explicit proof requires some messy calculation. The resulting module is called the *tensor induced module* and we denote it by

$$M^{\otimes(G/H)}.$$

It depends on the choice of the set $T$ of left coset representatives, but we shall see below that choosing a different such $T$ results in an isomorphic module.

## 5.2 Wreath products and the monomial representation

The ideas in the previous section are clarified if we introduce the so-called *wreath product*. Let $X$ be a set and let $S$ be a subgroup of $\mathcal{S}(X)$, the group of all permutations of $X$. Similarly, let $Y$ be another set and let $H$ be a subgroup of $\mathcal{S}(Y)$. The most important case will be $Y = H$ with $H$ imbedded in $\mathcal{S}(H)$ through left multiplication. Consider those $s \in \mathcal{S}(X \times Y)$ which permute the 'rows' $x \times Y$. For each such $s$, we have

$$s(x, y) = (\overline{s}(x), h_x(y)) \qquad \text{where} \quad \overline{s} \in \mathcal{S}(X) \quad \text{and each} \quad h_x \in \mathcal{S}(Y).$$

Let $\mathcal{W}$ be the subgroup of $\mathcal{S}(X \times Y)$ for which $\overline{s} \in S$ and $h_x \in H$ for each $x \in X$. $\mathcal{W}$ may be described as a semi-direct product of two subgroups. First, consider the subgroup of those $s' \in \mathcal{W}$ which send each row $x \times Y$ into itself by an element of $H$, i.e. $s'(x \times y) = x \times h_x(y)$ with $h_x \in H$. The correspondence $s' \leftrightarrow \prod h_x$ identifies this subgroup with $H^X$, the direct

## Wreath products and the monomial representation

product of $|X|$ copies of $H$ indexed by $X$. Second, consider the subgroup of all $s'' \in \mathcal{W}$ of the form $s \times \text{Id}$ for some $s \in S$. This subgroup may be identified with $S$, and its elements may be thought of as simply permuting the rows without changing them. It is clear that $S \cap H^X = \{1\}$. Moreover, $S$ normalizes $H^X$ since

$$s^{-1}\Big(\prod_{x \in X} h_x\Big)s = \prod_{x \in X} h_{s(x)}$$

for $s \in S$ and $\prod h_x \in H^X$. Finally, it is not hard to see that $\mathcal{W}$ is generated by $S$ and $H^X$, so we have $\mathcal{W} = H^X \rtimes S$. We call this group the *wreath product* and we denote it $S \int H$.

**Exercise 5.2.1.** Show that $\mathcal{W}$ is a subgroup of $\mathcal{S}(X \times Y)$.

*A note on notation.* The definition of wreath product commonly used in the literature is that of Hall (1959) where the assumption is made that groups act on sets on the right and that permutations compose to the right. In this monograph, we have found it convenient to treat permutations as functions and have them compose to the left. This is more consistent with the commonly used conventions in most developments of homological algebra. For that reason, our definition is slightly different from Hall's definition, and we have used the symbol $\int$ rather than the commonly used symbol $\wr$. (The idea was to reverse the orientation of the symbol, and $\int$ was the closest choice.)

Let $G$ be a group and $H$ a subgroup of finite index. As above, let $T$ be a set of left coset representatives of $H$ in $G$. Define the *monomial representation* $\Phi : G \to \mathcal{S}(G/H) \int H$ as follows. As above

$$gt = t_g h_{g,t} \qquad t_g \in T, h_{g,t} \in H,$$

and $t \mapsto t_g$ induces a permutation $s = \pi(g) \in \mathcal{S}(G/H)$. Identifying $T$ with $X = G/H$, define

$$\Phi(g) = \pi(g) \prod_{t \in T} h_{g,t}.$$

It is easy to check that $\Phi$ is a group monomorphism. Moreover, if we use a different set $T'$ of left coset representatives, the monomorphism $\Phi'$ defined from it differs from $\Phi$ by an inner automorphism of $\mathcal{S}(G/H) \int H$— i.e. $\Phi' = u\Phi u^{-1}$ for an appropriate $u \in \mathcal{S}(G/H) \int H$. As a consequence of this fact any functorial construction which is unaffected by such inner automorphisms does not depend on the set $T$.

Suppose now that $M$ is a $kH$-module. Consider as above $M^{\otimes X}$. Let $H^X$ act by

$$\Big(\prod h_x\Big) \otimes_{x \in X} m_x = \otimes_{x \in X} h_x m_x.$$

Similarly, let $S$ act by permuting the factors, i.e.

$$s(\otimes_{x\in X} m_x) = \otimes_{x\in X} m_{s^{-1}x}.$$

By being exceedingly careful, the reader may check that

$$\left(s\left(\prod_{x\in X} h_x\right)s^{-1}\right)\left(\otimes_{x\in X} m_x\right) = s\left(\prod_{x\in X} h_x\left(s^{-1}\left(\otimes_{x\in X} m_x\right)\right)\right)$$

so that we have an action of $S \wr H$ on $M^{\otimes X}$.

Let $G$ be a group, $H$ a subgroup of finite index, and $M$ a $kH$-module. Then as $k$-modules, we have $M^{\otimes G/H} \cong \bigotimes_{x \in X} M$ where $X = G/H$. It is not hard to see that the action of $G$ on $M^{\otimes G/H}$ defined previously factors through the homomorphism $\Phi : G \to \mathcal{S}(G/H) \wr H$.

**Proposition 5.2.1.** *Tensor induction has the following properties.*

*(a) (Transitivity) If $H$ is a subgroup of $K$ and $K$ is a subgroup of $G$, then*

$$(M^{\otimes K/H})^{\otimes G/K} \cong M^{\otimes G/H}.$$

*(b) $(M \otimes N)^{\otimes G/H} \cong M^{\otimes G/H} \otimes N^{\otimes G/H}$.*

*(c) Let $K$ and $H$ be subgroups of $G$ and let $G = \cup_{x \in D} KxH$ be a double coset decomposition. Then as a $K$-module obtained by restriction from the tensor induced $G$-module,*

$$(M^{\otimes G/H}) \cong \bigotimes_{x \in D} (x^*(M)^{\otimes K/(K \cap xHx^{-1})}).$$

**Proofs.** These are tedious but quite routine calculations. The reader will be excused if he or she accepts them on faith. See Evens (1963) for proofs. □

If we take $M = k \oplus A$ where $A$ is an $H$-module, then the tensor induced module from $M$ breaks up as a sum

$$k \oplus A{\uparrow}_H^G \oplus \cdots \oplus A^{\otimes G/H}$$

where the intermediate terms are induced from appropriate subgroups of $G$ in the ordinary sense. To see this let $M_i$ be the sum of those terms with $i$ tensor factors equal to $A$. Let $S'$ be the subgroup of $\mathcal{S}(G/H)$ which fixes the first $i$ elements of $X = G/H$. ($S' \cong S_i \times S_{(G:H)-i}$ where the first factor permutes the first $i$ terms and the second factor permutes the remaining terms.) It is possible to show that $M_i$ is the $\mathcal{S}(X) \wr H$-module induced from the $S' \wr H$-module $A^{\otimes i} \otimes k^{(G:H)-i}$. Since $\Phi$ is a monomorphism, we may view $G$ as imbedded in $\mathcal{S}(G/H)$, so the double coset formula shows us that $M_i$ as a $G$-module breaks up as a direct sum of submodules each of which is induced from an appropriate subgroup of $G$. For the case $i = 1$, one may check explicitly that the action of $G$ on $M_1$ (through $\Phi$) agrees with the action on the induced module $A\uparrow_H^G$ as described in Section 5.1. (Identify $1 \otimes \cdots \otimes a \otimes \cdots \otimes 1$ where $a$ is in the position indexed by $t$ with $t \otimes a$.)

**Exercise 5.2.2.** Verify the above assertions about the modules $M_i$.

In order to make use of the above constructions for complexes, we need to introduce appropriate signs because of the permutation of factors in the tensor product. Let $S \subseteq \mathcal{S}(X)$ as above, and let $U$ be a $kH$-complex (or even just a graded $kH$-module.) *Suppose $X$ is ordered.* For $s \in S$, define

$$s(\otimes_{x \in X} u_x) = (-1)^\epsilon \otimes_{x \in X} u_{s^{-1}x}$$

where

$$\epsilon = \sum_{x<y}(s(x) > s(y)) \deg u_x \deg u_y.$$

Here we adopt the computer programming convention that

$$(P) = \begin{cases} 1 & \text{if } P \text{ is true} \\ 0 & \text{if } P \text{ is false.} \end{cases}$$

It is tedious but routine to show that this defines an action which commutes with the action of $H^X$ and also with the differential in $U^{\otimes X}$. Note that this action depends on the ordering. This is also true in the ungraded case, but the isomorphisms which reorder the factors in a tensor product are essentially innocuous and usually can be ignored when no signs are involved. However, the reordering isomorphisms in the graded case will generally introduce signs. In particular, when we carry things back to a group $G$ via a monomial representation $\Phi$, we must remember that the action of $G$ on $U^{\otimes X} = U^{\otimes G/H}$ will depend on the ordering of the set T of left coset representatives. Fortunately, for anything involving only elements of even degree the sign will be $+1$.

## 5.3 Cohomology of wreath products

In this section, as elsewhere, we assume that $k$ is at least a Dedekind domain so that the Künneth formulas apply.

Let $S$ be a subgroup of $\mathcal{S}(X)$, and let $H$ be viewed as a subgroup of $\mathcal{S}(Y)$ with $Y = H$ as above. (We concentrate on that case, but the general theory is essentially the same.) Let $\eta : U \to k$ be a $kH$-projective resolution. Then $\eta$ induces a $k(H^X)$-projective resolution $\eta^{\otimes |X|} : U^{\otimes X} \to$

$k^{\otimes X} = k$. As discussed in the previous sections, there is an action (including signs) of $S$ on $U^{\otimes X}$ as a complex so that we are in the good situation for a semi-direct product as discussed in Section 2.5. Thus, if $\epsilon : W \to k$ is a $kS$-projective resolution, it follows that

$$\epsilon \otimes \eta^{\otimes |X|} : W \otimes U^{\otimes X} \to k$$

is a $k(S \wr H)$-projective resolution. Hence, if $M$ is a $kH$-module, then

$$H^*(S \wr H, M^{\otimes G/H})$$

may be calculated as the cohomology of the complex

$$\operatorname{Hom}_{S \wr H}(W \otimes (U^{\otimes X}), M^{\otimes X}) \cong \operatorname{Hom}_S(W, \operatorname{Hom}_{H^X}(U^{\otimes X}, M^{\otimes X}))$$
$$\cong \operatorname{Hom}_S(W, \operatorname{Hom}_H(U, M)^{\otimes X}).$$

Here we use quite extensively the general assumptions we have made, e.g. Künneth formulas apply, $U$ is finitely generated in each degree, etc. The reader should check the details. In particular, if $k$ is a field, we obtain an important theorem due to Nakaoka.

**Theorem 5.3.1** (Nakaoka). *Let $k$ be a field and suppose $X$ and $H$ are finite. Then*

$$H^*(S \wr H, k) \cong H^*(S, H^*(H, k)^{\otimes X}).$$

*Moreover, the isomorphism is an isomorphism of rings provided we employ the usual sign conventions in defining the product on the right.*

**Proof.** If $k$ is a field, we can choose $U \to k$ to be a *minimal* resolution so that the complex $\operatorname{Hom}_H(U, k)$ has trivial differential. It follows that

$$\operatorname{Hom}_H(U, k)^{\otimes X}$$

also has trivial differential. Thus the differential in the complex

$$\operatorname{Hom}_S(W, \operatorname{Hom}_H(U, k)^{\otimes X})$$

arises entirely from the differential in $W$, and the cohomology of this complex is

$$H^*(S, \operatorname{Hom}_H(U, k)^{\otimes X}) = H^*(S, H^*(H, k)^{\otimes X}),$$

since $\operatorname{Hom}_H(U, k) = H^*(H, k)$.

We leave it to the reader to prove the assertion about the product structure. □

Note that the ring structure on the right is simply the cup product for the cohomology of $S$ with coefficients in the $S$-ring appearing as a coefficient module—where appropriate signs are included to deal with the effect of the grading.

In the case that $k$ is a field, it follows from Nakaoka's theorem that $H^*(S \int H, k)$ contains the subring

$$H^0(S, H^*(H, k)^{\otimes X}) = (H^*(H, k)^{\otimes X})^S$$

and we can pick out certain invariants of the action of $S$ which play a specially important role. Namely, if $\alpha \in H^n(H, k)$ is of *even* degree, then

$$\alpha^{\otimes |X|} = \underbrace{\alpha \otimes \alpha \otimes \cdots \otimes \alpha}_{|X| \text{ times}} \in H^*(H, k)^{\otimes X}$$

is clearly such an invariant and we denote it $1 \int \alpha \in H^{n|X|}(S \int H, k)$.

We can define such a class even if $\alpha$ is not of even degree and/or $k$ is not a field, but the construction is a bit more difficult. We now show how to do it. The construction and most of what is done in this section is an adaptation of the construction by Steenrod of the *Steenrod reduced powers*. (See Steenrod (1962).)

Let $U \to k$ be any $kH$-resolution, and let $f : U_n \to k$ be a cocycle representing $\alpha \in H^n(H, k)$. We suppose for the moment that $n$ is even, but we will show how to drop that assumption later. Let $\epsilon : W \to k$ be the augmentation as above, and consider the cocycle

$$\epsilon \int f = \epsilon \otimes f^{\otimes |X|} : W \otimes U^{\otimes X} \to k \otimes k^{\otimes X} = k.$$

We shall denote by $1 \int \alpha$ the class in $H^{n|X|}(S \int H, k)$ represented by $\epsilon \int f$.

Note that $\epsilon \otimes \mathrm{Id}$ induces a map

$$\mathrm{Hom}_{S \int H}(k \otimes U^{\otimes X}, k) \to \mathrm{Hom}_{S \int H}(W \otimes U^{\otimes X}, k).$$

The left hand side may be identified with

$$(\mathrm{Hom}_H(U, k)^{\otimes X})^S$$

and $f^{\otimes |X|}$ is a cocycle in that complex. Its image in the cohomology of the right hand side is the desired class. In the case that $k$ is a field the class on the left would be exactly what we wanted. Notice, however, that

$$H^*((\mathrm{Hom}_H(U, k)^{\otimes X})^S) \neq (H^*(\mathrm{Hom}_H(U, k)^{\otimes X})^S$$

in general which complicates the construction in the general case.

So far the construction has been fairly straightforward except perhaps for the cumbersome notation. The tricky point is to show that $1 \int \alpha$ is *well defined*, i.e. *it depends only on $\alpha$ and not on the cocycle $f$ representing $\alpha$*. This point often leads people astray. The desired independence follows from the following lemma.

**Lemma 5.3.2** (Steenrod). *Let $F, G : U \to L$ be chain homotopic maps of $kH$-complexes. Then*

$$\mathrm{Id} \otimes F^{\otimes |X|}, \ \mathrm{Id} \otimes G^{\otimes |X|} : W \otimes U^{\otimes X} \to W \otimes L^{\otimes X}$$

*are chain homotopic maps of $k(S \int H)$-complexes.*

To show that $1 \int \alpha$ is well defined, we use the lemma as follows. Let $f$ and $g$ both represent $\alpha \in H^n(H, k)$. Let $L_n = k$ and suppose that $L_r = 0$ for $r \neq n$. Then $f$ and $g$ may be viewed as chain homotopic maps of chain complexes $U \to L$ of degree 0. It follows that $\mathrm{Id} \otimes f^{\otimes |X|}$ and $\mathrm{Id} \otimes g^{\otimes |X|}$ are chain homotopic, so $(\epsilon \otimes \mathrm{Id}) \circ (\mathrm{Id} \otimes f^{\otimes |X|}) = \epsilon \otimes f^{\otimes |X|}$ and $(\epsilon \otimes \mathrm{Id}) \circ (\mathrm{Id} \otimes g^{\otimes |X|}) = \epsilon \otimes g^{\otimes |X|}$ are chain homotopic maps of $W \otimes U^{\otimes X} \to L^{\otimes X}$. However, the right hand side has only one component which is in degree $n|X|$ and is isomorphic to $k$. It follows that we can identify these maps with $\epsilon \int f$ and $\epsilon \int g$, and to say they are chain homotopic is simply to say they are cohomologous since the differential in $L$ is trivial.

**Proof.** To say $F$ and $G$ are chain homotopic is to say that there is a map $P : U \to L$ of degree 1 such that

$$G - F = dP + Pd.$$

This statement may be translated (using a geometric analogy) as follows. Let $I$ be the complex with $I_0 = kp_0 \oplus kp_1$, $I_1 = ke$, and $de = p_1 - p_0$. $I$ is the chain complex of the unit interval with two vertices and one edge. (Note that with the augmentation $\rho : I \to k$ defined by $\rho : p_0, p_1 \mapsto 1$, $I$ may also be viewed as a $kN$-projective resolution of $k$ for $N$ the trivial group.) Given a chain homotopy as above, we may define a (degree 0) $kH$-map of chain complexes

$$\tilde{P} : I \otimes U \to L$$

by

$$\tilde{P}(p_0 \otimes u) = F(u)$$
$$\tilde{P}(p_1 \otimes u) = G(u)$$
$$\tilde{P}(e \otimes u) = P(u).$$

It is easy to check that the chain homotopy relation is equivalent to saying that $\tilde{P}$ commutes with the differentials. Conversely, given any such $\tilde{P}$, we may use it to define $F, G,$ and $P$.

To prove the lemma, we need to construct a $k(S \int H)$-map

$$\tilde{Q} : I \otimes W \otimes U^{\otimes X} \to W \otimes L^{\otimes X}$$

which on $p_0 \otimes W \otimes U^{\otimes X}$ is $F^{\otimes |X|}$ and on $p_1 \otimes W \otimes U^{\otimes X}$ is $G^{\otimes |X|}$. We do this by first constructing an $S$-map

$$\tilde{J} : I \otimes W \to I^{\otimes X} \otimes W$$

such that

$$\tilde{J}(p_0 \otimes w) = p_0^{\otimes |X|} \otimes w$$
$$\tilde{J}(p_1 \otimes w) = p_1^{\otimes |X|} \otimes w.$$

Then $\tilde{Q}$ is the composition

$$I \otimes W \otimes U^{\otimes X} \to I^{\otimes X} \otimes W \otimes U^{\otimes X}$$
$$\cong W \otimes (I^{\otimes X} \otimes U^{\otimes X}) \cong W \otimes (I \otimes U)^{\otimes X} \to W \otimes L^{\otimes X}$$

where the first map is $\tilde{J} \otimes \mathrm{Id}$, the next two are changes of order, and the last one is $\mathrm{Id} \otimes \tilde{P}^{\otimes |X|}$. This composition carries $p_0 \otimes w \otimes \otimes_{x \in X} u_x$ to $p_0^{\otimes |X|} \otimes w \otimes \otimes_{x \in X} u_x$ then to $w \otimes \otimes_x (p_0 \otimes u_x)$ and then to $w \otimes \otimes_x F(u_x)$ and similarly for $p_1$ as required. (Notice that we don't even have to worry about sign changes since each twisting involves an element of even degree.) It is easy to see that this is a $k(S \int H)$-map.

To construct the desired $\tilde{J}$ we argue as follows. Since $I \to k$ is a projective resolution for the trivial group and $k(1 \times S) = kS$, $I \otimes W \to k$ is a $kS$-projective resolution. Also, since $k(1^X \ltimes S) = kS$, $I^{\otimes X} \otimes W \to k$ is also a $kS$-projective resolution. Hence, we know there is a $kS$ map from one to the other, but we want to construct one with certain special properties. Note that $I_0 \otimes W$ is a $kS$-subcomplex of $I \otimes W$, and, using the desired formulas for $\tilde{J}$, we can define a $kS$-map

$$\tilde{J}' : I_0 \otimes W \to I^{\otimes X} \otimes W$$

consistent with the augmentations. The trick is to show that this can be *extended* to the desired map $\tilde{J} : I \otimes W \to I^{\otimes X} \otimes W$. Let $A = I \otimes W$ and let $B = I^{\otimes X} \otimes W$. Write $A = A' \oplus A''$ where $A' = I_0 \otimes W$ and $A'' = I_1 \otimes W$ is not a subcomplex. We define $\tilde{J}$ by induction on the degree $s$ in the complex $A$. For $s = 0$, $A'_0 = A_0$, so no extension is necessary. Suppose it has been defined for degree $s$. In addition, since $A''_{s+1} \to A_s \to A_{s-1}$ is trivial, the composition

$$\begin{array}{c} A''_{s+1} \\ \downarrow \\ A_s \longrightarrow B_s \\ \downarrow \\ B_{s-1} \end{array}$$

is also trivial, i.e. $\tilde{J}_s(d(A''_{s+1})) \subseteq d(B_{s+1})$. Since $A''_{s+1}$ is $kS$-projective, we can define $\tilde{J}''_{s+1} : A''_{s+1} \to B_{s+1}$ making

$$\begin{array}{ccc} A''_{s+1} & \longrightarrow & B_{s+1} \\ \downarrow & & \downarrow \\ A_s & \longrightarrow & B_s \end{array}$$

commute. Then $\tilde{J}_{s+1} = \tilde{J}'_{s+1} \oplus \tilde{J}''_{s+1}$ clearly extends the map of complexes one step further. □

The above argument shows that $1 \int \alpha$ is independent of $f$, and by extension of the argument it is not hard to see that it is independent of the $kH$-resolution $U \to k$. It is even easier to see that it is independent of $W$. For, suppose $\epsilon' : W' \to k$ and $\epsilon : W \to k$ are two $kS$-projective resolutions. Then there is a $kS$-map $W' \to W$ consistent with the augmentations and clearly $\epsilon \otimes f^{\otimes |X|}$ is carried into $\epsilon' \otimes f^{\otimes |X|}$ by this map so that they define the same cohomology class in $H^*(S \int H, k)$. This argument can even be generalized to the case that $S'$ is a subgroup of $S$, and $\epsilon' : W' \to k$ is a $kS'$-projective resolution. This shows that $1 \int \alpha$ is independent of the subgroup $S$ in the sense that classes defined for different $S$ are coherently related to one another; in particular, they are all restrictions of the class defined in $H^*(S(X) \int H, k)$.

## 5.4 Odd degree and other variations on the theme

In defining $1 \int \alpha$ we assumed $\alpha$ was of even degree to avoid sign problems. If $\alpha \in H^n(H, k)$ with $n$ odd, then we have to modify the discussion as follows. Let $S \leq S(X)$ act on

$$\tilde{k} = \underbrace{k \otimes k \otimes \cdots \otimes k}_{|X| \text{ times}}$$

by permuting the factors *and also multiplying by the sign of the permutation*. Of course, as a $k(H^X)$-module, $\tilde{k} \cong k$. Also, if $S$ consists only of even permutations, $\tilde{k} \cong k$ even as a $k(S \int H)$-module. (That would be the case, for example, if $S$ were generated by a cycle of odd length, e.g. $(12 \ldots p) \in S(p)$ for $p$ odd.) In any case, for $f \in \text{Hom}_H(U_n, k)$ representing $\alpha$,

$$\epsilon \otimes f^{\otimes |X|} : W \otimes U^{\otimes X} \to \tilde{k}$$

is a $k(S \int H)$-morphism and as such defines a class

$$1 \int \alpha \in H^{n|X|}(S \int H, \tilde{k}).$$

In addition, if we look at the proof that this class is well defined in the case of even degree, we can see that with the defined action (including signs) $L^{\otimes X}$ in its single non-zero degree is isomorphic as an $S$-module to $\tilde{k}$. (Note also that Steenrod's Lemma (Lemma 5.3.2) does not depend on evenness or oddness since it works for any map of chain complexes.)

For most of what we shall be interested in, the sign is not an issue. This includes the following cases:
(a) $n$ is even;
(b) $n$ is odd, $k$ is a field of odd characteristic $p$; and $S$ is generated by a cycle of length $p$,
(c) $k$ is a field of characteristic 2 in which case signs do not matter anyway.

It is also possible to define $1 \int \alpha$ for coefficient modules $M$ other than $k$. This was done for even degrees in Evens (1963). In this case, if $\alpha \in H^n(H, M)$, then $1 \int \alpha \in H^{n|X|}(S \int H, M^{\otimes X})$.

If $\alpha \in H^*(H, k)$ is not homogeneous but is instead a *sum* of elements of even degree, we can use essentially the same construction to define an inhomogeneous class $1 \int \alpha \in H^*(S \int H, k)$. In particular, let $\alpha$ and $\beta$ be homogeneous elements of *possibly different* even degrees $n$ and $m$, and suppose $f$ is a cocycle representing $\alpha$ and $g$ is a cocycle representing $\beta$. Then as before define $1 \int (\alpha + \beta)$ to be the cohomology class represented by

$$\epsilon \otimes (f + g)^{\otimes q}$$

where $q = |X|$. By expanding this out, we obtain a sum of terms of the form

$$\epsilon \otimes (h_{j_1} \otimes h_{j_2} \otimes \cdots \otimes h_{j_q})$$

where each $h_j = f$ or $g$. The term where all $h_j = f$ is $\epsilon \otimes f^{\otimes q}$, which represents $1 \int \alpha$, and similarly the term with all $h_j = g$ represents $1 \int \beta$. The remaining terms can be divided into $S$-orbits $\mathcal{J}$. Let $S_{\mathcal{J}}$ be the isotropy subgroup of a term

$$T = \epsilon \otimes (h_{j_1} \otimes h_{j_2} \otimes \cdots \otimes h_{j_q})$$

in the orbit $\mathcal{J}$. Note that $S_{\mathcal{J}} \leq S_r \times S_s$ where $S_r$ permutes $r$ factors in $T$ equal $f$ and $S_s$ permutes $s$ factors equal $g$, $r + s = q$. In particular, $S_{\mathcal{J}}$ does not act transitively on $X$. It is not hard to see that the sum of the terms in the orbit $\mathcal{J}$ represents

$$\mathrm{cor}_{[S_{\mathcal{J}} \int H \to S \int H]}(\tau)$$

where $\tau \in H^*(S_{\mathcal{J}} \int H, k)$ is represented by $T$.

**Proposition 5.4.1.** Let $\alpha, \beta \in H^*(H,k)$ of even degree. Then we have (in $H^*(S \int H, k)$)

$$1 \int (\alpha + \beta) = 1 \int \alpha + \sum_{\mathcal{J}} \mathrm{cor}_{[S_{\mathcal{J}} \int H \to S \int H]}(\tau_{\mathcal{J}}) + 1 \int \beta.$$

In particular, if $S$ is cyclic of prime order $p = |X|$, we have

$$1 \int (\alpha + \beta) = 1 \int \alpha + \mathrm{cor}_{[H^p \to S \int H]}(\tau) + 1 \int \beta$$

for some $\tau \in H^*(H^p, k)$.

**Proof.** For the prime case, note that $S \cap (S_r \times S_s) = \{1\}$ if $r + s = p$ and neither $r$ nor $s$ is 0. Hence, all the subgroups $S_{\mathcal{J}} \int H = H^p$, and the corestrictions can be combined into one term. □

One interesting case is $\alpha = 1 \in k = H^0(H, k)$ and $\beta$ of positive even degree. Note, however, that in the proposition $\alpha$ and $\beta$ could be of the same degree or even both inhomogeneous of mixed degree.

# 6
# The norm map

## 6.1 Definition of the norm map

Let $G$ be a group and $H$ a subgroup of finite index. Let $\Phi : G \to S \wr H$ be the monomial representation defined in Section 5.2. Here we can take $S = \mathcal{S}(G/H)$ or any subgroup of $\mathcal{S}(G/H)$ which contains the image of $\pi : G \to \mathcal{S}(G/H)$ (the representation of $G$ as a group of permutations of its left coset space $G/H$). In particular, if $H$ is normal in $G$, then that image is isomorphic to the factor group $G/H$ itself.

If $\alpha \in H^*(H, k)$ is of even degree (but as in Section 5.4 not necessarily homogeneous), define

$$N_{H \to G}(\alpha) = \Phi^*(1 \wr \alpha) \in H^*(G, k).$$

If $\alpha$ is homogeneous of degree $n$, then $N_{H \to G}(\alpha) \in H^{n(G:H)}(G, k)$. This cohomology class is well defined, since if we were to use another monomial representation $\Phi'$ associated with another set $T'$ of left coset representatives, $\Phi' = \iota_u \Phi$ for an appropriate inner automorphism $\iota_u$ of $S \wr H$, so $\Phi'^*(1 \wr \alpha) = \Phi^*(\iota_u^*(1 \wr \alpha)) = \Phi^*(1 \wr \alpha)$. Clearly, $N_{G \to G} = \mathrm{Id}$.

**Theorem 6.1.1.** *Suppose $H$ is a subgroup of $G$ of finite index. $N_{H \to G}$ has the following properties.*

*(N1) If $H$ is a subgroup of $K$ and $K$ is a subgroup of $G$, then for $\alpha \in H^*(H, k)$ of even degree,*

$$N_{K \to G}(N_{H \to K}(\alpha)) = N_{H \to G}(\alpha).$$

*(N2) If $\alpha, \beta \in H^*(H, k)$ are of even degree, then*

$$N_{H \to G}(\alpha\beta) = N_{H \to G}(\alpha) N_{H \to G}(\beta).$$

*(N3) If $G = \bigcup_{x \in D} KxH$ is a double coset decomposition of $G$, then for $\alpha \in H^*(H, k)$ of even degree,*

$$\mathrm{res}_{G \to K}(N_{H \to G}(\alpha)) = \prod_{x \in D} N_{K \cap xHx^{-1} \to K}(\mathrm{res}_{xHx^{-1} \to K \cap xHx^{-1}}(x^*\alpha))$$

*where $\prod$ denotes the cup product operation and the order is irrelevant since everything is of even degree.*

(N4) If $H$ is normal in $G$, then
$$\mathrm{res}_{G \to H}(N_{H \to G}(\alpha)) = \prod_{y \in G/H} y^*(\alpha).$$

(N5) Let $H'$ be a subgroup of $G'$ and let $H$ be a subgroup of $G$, in both cases of finite index. Suppose $\phi : G' \to G$ is a homomorphism such that $\phi(H') \subseteq H$ and $\phi$ induces a one-to-one correspondence $G'/H' \cong G/H$ of coset spaces. Let $\phi'$ denote the restriction of $\phi$ to $H'$. Then
$$N_{H' \to G'}(\phi'^*(\alpha)) = \phi^*(N_{H \to G}(\alpha))$$
for $\alpha \in H^*(H, k)$ of even degree.

**Note.** If $\alpha \in H^0(H, k) = k$, $N_{H \to G}(\alpha) = \alpha^{(G:H)}$, and so in that case we have the formula
$$N_{H \to G}(\mathrm{res}_{G \to H}(\alpha)) = \alpha^{(G:H)}.$$
However, if $\alpha$ has degree $> 0$, this formula is generally false. Also, the more general formula we might expect for $N_{H \to G}(\mathrm{res}_{G \to H}(\alpha)\beta)$ is false. (See Evens (1963, Section 6, pp. 62–3 for details.)) The corresponding formulas for corestriction are of course true.

**Corollary 6.1.2.** *Let $G$ be a finite group and $H$ a non-trivial subgroup. Assume $k = \mathbf{Z}$ or $k$ is a field of characteristic $p$ where $p \mid |H|$. Then $\mathrm{res}_{G \to H}$ is non-trivial in infinitely many positive degrees.*

**Proof.** Let $p$ be a prime dividing $|H|$ and suppose $p = \mathrm{char}\, k$ if $k$ is a field. Under the given hypotheses, $H$ contains a *cyclic* subgroup of order $p$, and by the transitivity of restriction we may assume $H$ is that subgroup. Let $\alpha$ generate $H^2(H, k)$. By the double coset rule (Theorem 6.1.1),
$$\mathrm{res}_{G \to H}(N_{H \to G}(1 + \alpha))$$
$$= \prod_{x \in D} N_{xHx^{-1} \cap H \to H}(\mathrm{res}_{xHx^{-1} \to xHx^{-1} \cap H}\, x^*(1 + \alpha))$$
$$= \prod_{x \in N_G(H)/H} (1 + x^*(\alpha))$$
since $xHx^{-1} \cap H = \{1\}$ for all $x \notin N_G(H)$. However, since $H$ is cyclic, for each $x \in N_G(H)$, $x^*(\alpha) = r\alpha$ for some $r \in k$, and it is not hard to see that $r$ is a unit in $k$ mod $|H|$. It follows that $\exists \gamma \in H^{2(N_G(H):H)}(G, k)$ (some component of $N_{H \to G}(1 + \alpha)$) such that $\mathrm{res}(\gamma) = q\alpha^{(N_G(H):H)}$ where $q$ is also a unit mod $|H|$. Hence, $\gamma$ and its powers all restrict non-trivially in $H^*(H, k)$. □

We can now state the following elementary result—which as far as we know cannot be proved without the above argument or its equivalent. (There are also relatively simple topological arguments but they use substantial facts from algebraic topology.)

**Corollary 6.1.3.** *If $G$ is a non-trivial finite group, then $H^n(G, \mathbf{Z}) \neq 0$ for infinitely many $n > 0$. In fact, the result holds for each p-primary component $H^*(G, \mathbf{Z}; p)$ for which $p \mid |G|$.*

*Note.* One often encourages topological arguments in the literature which use the highest *Chern class* of a unitary representation induced from a character $\chi \in \mathrm{Hom}(H, \mathbf{C}^*)$ of a subgroup. Since the highest Chern class in this case is just the norm of the corresponding element of $H^2(H, \mathbf{Z})$, such arguments usually mask an implicit use of the norm. It is usually more direct just to use the norm. (See Evens (1965).)

**Theorem 6.1.4** (Additivity Rules). *Suppose $H$ is a subgroup of $G$ of finite index, and $\alpha, \beta \in H^*(G, k)$ are homogeneous elements of even degree.*
 (A1) *If $n = \deg \alpha > 0$, then*
$$N_{H \to G}(1 + \alpha) = 1 + \mathrm{cor}_{H \to G}(\alpha) + \cdots + N_{H \to G}(\alpha)$$
*where the intermediate terms are corestrictions with degrees between $n$ and $n(G : H)$.*
 (A2) *In general,*
$$N_{H \to G}(\alpha + \beta) = N_{H \to G}(\alpha) + \cdots + N_{H \to G}(\beta)$$
*where the intermediate terms are corestrictions from proper subgroups of $G$. If $H$ is normal in $G$, these subgroups contain $H$.*
 (A3) *Suppose $H$ is normal in $G$ of prime index $p$. Then*
$$N_{H \to G}(\alpha + \beta) = {}'N_{H \to G}(\alpha) + \mathrm{cor}_{H \to G}(\mu) + N_{H \to G}(\beta)$$
*for some $\mu \in H^*(H, k)$. If $\deg \alpha = n < \deg \beta = m$, then $\mu$ is a sum of terms of degrees strictly between $pn$ and $pm$. If $n = m$, then $\mu$ has degree $pn$.*

## 6.2 Proofs of the properties of the norm

Property (N1). Let $U \to k$ be a $kH$-projective resolution, and let $\epsilon' : W' \to k$ be a $kS(K/H)$-projective resolution. Then, through the appropriate map $\Phi' : K \to S(K/H) \int H$, we may view $W' \otimes U^{\otimes K/H} \to k$ as a $kK$-projective resolution. If we do that $N_{H \to K}(\alpha)$ is represented by
$$\epsilon' \otimes f^{\otimes |K/H|}.$$

Let $\epsilon'' : W'' \to k$ be a $kS(G/K)$-projective resolution. Then through the appropriate map $\Phi'' : G \to S(G/K) \int K$ we may view
$$W'' \otimes (W' \otimes U^{\otimes K/H})^{\otimes G/K} \cong (W'' \otimes W'^{\otimes G/K}) \otimes U^{\otimes G/H} \to k$$

as a $kG$-projective resolution. If we do so, $N_{K \to G}(N_{H \to K}(\alpha))$ is represented by

$$\epsilon'' \otimes (\epsilon' \otimes f^{\otimes |K/H|})^{\otimes |G/K|} \cong (\epsilon'' \otimes \epsilon'^{\otimes |G/K|}) \otimes f^{\otimes |G/H|}.$$

On the other hand, it is not hard to see that this action of $G$ factors through an appropriate map

$$\Phi'' : G \to (\mathcal{S}(G/K) \int \mathcal{S}(K/H)) \int H, \leq \mathcal{S}(G/H) \int H$$

so we can identify the cocycle on the right as representing

$$1 \int \alpha \in H^*((\mathcal{S}(G/K) \int \mathcal{S}(K/H)) \int H, k)$$

and carry it back to $G$ as called for.

Note that we have used the identification $S'' \int (S' \int H) \cong (S'' \int S') \int H$ where $S''$ acts on $X''$, $S'$ acts on $X'$, and $H$ acts on $Y$. In that case both sides act on $X'' \times X' \times Y$.

Property (N2). This follows if we can prove the formula

$$1 \int (\alpha \beta) = (1 \int \alpha)(1 \int \beta)$$

or the corresponding formula for the external product $\alpha \times \beta$. It is left to the reader to verify these formulas.

Property (N3). Let $T(x)$ be a set of left coset representatives of $H_x = K \cap xHx^{-1}$ in $K$ so that $T = \cup_{x \in D} T(x)x$ is a set of left coset representatives of $H$ in $G$. Let $\Phi_x : K \to \mathcal{S}(K/H_x) \int H_x = S_x \int H_x$ be the monomial representation given by $T(x)$. Then

$$ktx = t'h'_{k,t}x = t'xx^{-1}h'_{k,t}x \qquad \text{for } k \in K, t, t' \in T(x), h'_{k,t} \in K \cap xHx^{-1}$$

shows that we have a commutative diagram

$$
\begin{array}{ccc}
G & \xrightarrow{\Phi} & \mathcal{S}(G/H) \int H \\
\uparrow & & \uparrow \\
K & & (\prod_{x \in D} S_x) \int H \\
{\scriptstyle \Delta} \searrow & & \uparrow B \\
& \prod_{x \in D} K & \prod_{x \in D}(S_x \int H) \\
& \prod_{x \in D} \Phi_x \searrow & \uparrow C \\
& & \prod_{x \in D}(S_x \int H_x)
\end{array}
$$

where $\Delta$ is the $|D|$-fold diagonal homomorphism, $C$ is induced by a product of conjugations followed by inclusions, and $B$ is an appropriate map

## Proofs of the properties of the norm

(discussed below). To prove (N3), choose a resolution $U \to k$ for $H$, a resolution $W \to k$ for $\mathcal{S}(G/H)$, and resolutions $U(x) \to k$ and $W(x) \to k$ for $H_x$ and $S_x$ respectively for each $x \in D$. Starting with $\epsilon \otimes f^{\otimes |T|}$ on the upper right, follow it down through the diagram putting in the appropriate resolutions at each stage and using the appropriate maps between them.

The map $B$ is the extension to many factors of the twisting isomorphism $(S' \int H) \times (S'' \int H) \cong (S' \times S'') \int H$ which arises from the identification $(X' \times Y) \cup (X'' \times Y) = (X' \cup X'') \times Y$ for two disjoint sets $X'$ and $X''$. ($S' \times S''$ acts on $X' \cup X''$ by letting each factor act on its component: $(s' \times s'')x = s'x$ or $s''x$ according to $x \in X'$ or $x \in S''$.)

Property (N4). This follows immediately from (N3) once it is realized that a double coset decomposition is the same as a single coset decomposition for a normal subgroup.

Property (N5). This follows from the fact that $1 \int \alpha$ is natural with respect to the group homomorphism $S \int H' \to S \int H$ induced by $\phi' : H' \to H$. The rigid assumptions on $G', H', G$, and $H$ are necessary so that we can identify $G'/H'$ with $G/H$. We leave it to the reader to construct the appropriate diagrams.

Properties (A1) and (A2). Since the monomial representation $\Phi : G \to S \int H$ is a monomorphism, we may view $G$ as imbedded as a subgroup of $S \int H$, with $S = \mathcal{S}(G/H)$. We may now use Proposition 5.4.1. The two components of

$$1 \int (\alpha + \beta) = 1 \int \alpha + \cdots + 1 \int \beta$$

on the ends restrict to $N_{H \to G}(\alpha)$ and $N_{H \to G}(\beta)$ respectively. To calculate what happens to the intermediate terms under restriction, use the double coset rule to calculate

$$\mathrm{res}_{S \int H \to G} \circ \mathrm{cor}_{S_\mathcal{J} \int H \to S \int H}.$$

This yields a sum of corestrictions from subgroups of the form

$$G \cap y(S_\mathcal{J} \int H)y^{-1}.$$

However, it is easy to see that for $y \in S \int H$, any $y(S_\mathcal{J} \int H)y^{-1}$ is of the form $S'_\mathcal{J} \int H$ where $S'_\mathcal{J}$ is the isotropy group for some other term $T'$ in the orbit $\mathcal{J}$. In either case, these isotropy groups are of the form $S_r \times S_s$ where $r + s = (G : H)$. Here, $S_r$ permutes some subset of $G/H$ of size $r$, and $S_s$ permutes the rest. If $(S_r \times S_s) \int H \geq G$, then the image $\pi(G) \leq S_r \times S_s$, which is impossible since the latter does not act transitively on $G/H$. It follows that the intermediate terms involve corestrictions from proper subgroups.

To derive the explicit formula in (A1), note that for the first intermediate term, there is only one orbit, that of the term

$$\epsilon \otimes (f \otimes g \otimes \cdots \otimes g)$$

where $f$ represents $\alpha$ and $g$ represents 1. In that case, it is not hard to see that

$$G \cap S_J \int H = H.$$

Since $\mathrm{res}_{S_J \int H \to H}$ carries the above term to $\alpha$, the desired conclusion follows.

Property (A3). This follows from the second part of Proposition 5.4.1 since in that case $(G : H) = (S \int H : H^p) = p$.

## 6.3 The norm map for elementary abelian $p$-groups

Let $G = P \times P$ where $P$ is a cyclic group of prime order $p$. Since $H^n(P, \mathbf{Z}) = 0$ for $n$ odd, the Künneth Formula (as in Section 2.5) tells us that

$$H^{2n}(G, \mathbf{Z}) \cong \bigoplus_{i+j=n} H^{2i}(P, \mathbf{Z}) \otimes H^{2j}(P, \mathbf{Z})$$

since every Tor term will involve an odd and even degree term. Moreover, $\alpha \otimes \beta$ on the right corresponds to $\alpha \times \beta$ on the left, so $H^{2n}(G, \mathbf{Z})$ is spanned by the products $\alpha \times \beta$. Let $\chi$ generate $H^2(P, \mathbf{Z})$ (so $H^*(P, \mathbf{Z}) = \mathbf{Z}[\chi \,|\, p\chi = 0])$. It follows that any element of $H^{2p}(G, \mathbf{Z})$ can be written uniquely:

$$\sum_{i+j=p} a_{i,j} \chi^i \times \chi^j.$$

Identify $P$ with the subgroup $\{1\} \times P$ of $P \times P$.

**Proposition 6.3.1.** *With the above notation,*

$$N_{P \to P \times P}(\chi) = 1 \times \chi^p - \chi^{p-1} \times \chi.$$

**Proof.** Write $F = P = \{1\} \times P$, $E = P \times P \cong \mathbf{F}_p{}^2$, and $\xi = \chi \times 1, \eta = 1 \times \chi \in H^2(E, \mathbf{Z})$. As in Exercise 4.2.3, $H^2(E, \mathbf{Z}) \cong \mathrm{Hom}(E, \mathbf{Q}/\mathbf{Z}) = \hat{E}$, and since $E$ is elementary, the latter may be viewed as an $\mathbf{F}_p$-vector space with basis $\{\xi, \eta\}$. In what follows, we shall view $\xi$ and $\eta$ as homomorphisms $E \to \mathbf{Q}/\mathbf{Z}$ (with images contained in fact in $(1/p)\mathbf{Z}/\mathbf{Z} \cong \mathbf{F}_p$.) Then $F = \mathrm{Ker}\,\xi$, but we may also imbed it in $E$ as $F_i = \mathrm{Ker}(\eta - i\xi)$ for $i = 0, \ldots, p-1$. (Think of $F_i$ as the line $\eta - i\xi = 0$ and $F = F_\infty$ as the line $\xi = 0$.) We want to calculate the homogeneous polynomial of degree $p$

$$N_{F \to E}(\chi) = \sum_{j=0}^{p} a_j \eta^j \xi^{p-j}.$$

By the double coset formula, if we restrict $N_{F\to E}(\chi)$ to $F_i$, we obtain zero. (For $E = FF_i$ consists of *one* double coset and $F \cap F_i = \{0\}$.) However, we saw in Section 3.5 that the subring $S_{\mathbf{Z}}(\hat{E})$ generated by elements of degree 2 is in positive degrees just the symmetric algebra over $\mathbf{F}_p$ of the $\mathbf{F}_p$-vector space $\hat{E}$. Also, on this subring of $H^*(E, \mathbf{Z})$, restriction to $F_i$ is induced by the map of dual spaces $\hat{E} \to \hat{F}_i$. By calculating with symmetric algebras, it is not hard to see that $\operatorname{Ker} \operatorname{res}_{E \to F_i} \cap S_{\mathbf{Z}}(\hat{E})$ is the principal ideal generated by $\eta - i\xi$. It follows that $N_{F\to E}(\chi)$ is divisible in $S_{\mathbf{Z}}(\hat{E})$ by $\eta - i\xi$ for each $i$. Hence, it is divisible by their product, which is

$$\prod_{i=0}^{p-1}(\eta - i\xi) = \eta^p - \xi^{p-1}\eta,$$

so $N_{F\to E}(\chi) = c(\eta^p - \xi^{p-1}\eta)$ for some integer $c$. On the other hand, by the double coset formula for $\operatorname{res}_{E\to F} \circ N_{F\to E}$, restricting to $F$ (where $\eta \mapsto \chi, \xi \mapsto 0$) sends $N_{F\to E}(\chi)$ to $\chi^p$ so $c \equiv 1 \bmod p$. □

We will need the following result in order to prove an important theorem of Serre.

**Corollary 6.3.2.** *Let $E$ be an elementary abelian $p$-group and let $F$ be a subgroup. Then for each $\chi \in H^2(F, \mathbf{Z})$ we have*

$$N_{F\to E}(\chi) = \prod_{\operatorname{res}_{E\to F}(\eta)=\chi} \eta.$$

**Proof.** Suppose $E \geq E' \geq F$, and suppose the corollary has been established for the intermediate stages. Then

$$N_{F\to E}(\chi) = N_{E'\to E}(N_{F\to E'}(\chi)) = N_{E'\to E}\Big(\prod_{\eta' \mapsto \chi} \eta'\Big) = \prod_{\eta' \mapsto \chi} N_{E'\to E}(\eta')$$
$$= \prod_{\eta' \mapsto \chi}\Big(\prod_{\eta \mapsto \eta'} \eta\Big) = \prod_{\eta \mapsto \chi} \eta.$$

It follows that it suffices to prove the Corollary for $(E : F) = p$.

If $\chi = 0$, then $N(\chi) = 0$, so we may assume $\chi \neq 0$. Identify $\chi$ with an element of $\operatorname{Hom}(F, \mathbf{Q}/\mathbf{Z}) = H^2(F, \mathbf{Z})$ as above. Since $F$ is elementary abelian, it follows that the image of $\chi$ has order $p$ and $F_1 = \operatorname{Ker} \chi$ is of index $p$. Also, $\chi = \inf_{F/F_1 \to F} \chi_1$ for an appropriate $\chi_1 \in H^2(F/F_1, \mathbf{Z})$. Since $(E : F) = (E/F_1 : F/F_1) = p$, we have by Theorem 6.1.1(N5)

$$N_{F\to E}(\chi) = N_{F\to E}(\inf_{F/F_1 \to F} \chi_1)$$
$$= \inf_{E/F_1 \to E}(N_{F/F_1 \to E/F_1} \chi_1).$$

Also, there are exactly $p = (E : F)$ elements $\eta \in H^2(E, \mathbf{Z})$ such that $\mathrm{res}_{E \to F}\, \eta = \chi = \inf_{F/F_1 \to F}(\chi_1)$, and since there are $p$ elements of the form $\inf_{E/F_1 \to E}(\eta_1)$, where $\mathrm{res}_{E/F_1 \to F/F_1}(\eta_1) = \chi_1$, these are all the elements entering into the desired product. This reduces us to the case $E/F_1 > F/F_1$, i.e., it suffices to prove the corollary for $|E| = p^2$ and $|F| = p$. However, this is the case just considered in Proposition 6.3.1 where we obtained

$$\eta^p - \xi^{p-1}\eta = \prod_{i=0}^{p-1}(\eta - i\xi),$$

and the elements $\eta - i\xi$ are the $p$ elements of $H^2(E, \mathbf{Z})$ which restrict to $\chi$. □

**Proposition 6.3.3.** *Let $G$ be a group and $P$ a cyclic group of prime order $p$ and $k$ a field of characteristic $p$. Identify $G$ with the subgroup $1 \times G$ of $P \times G$. The map $N_{G \to P \times G} : H^{ev}(G, k) \to H^{ev}(P \times G, k)$ is a ring homomorphism. (For $p = 2$, it is not necessary to restrict to even degrees.)*

**Proof.** The norm is always multiplicative. In this case, it is also additive. For we have

$$N(\alpha + \beta) = N(\alpha) + \mathrm{cor}(\mu) + N(\beta)$$

by Theorem 6.1.4(A3). The result follows from the next lemma.

**Lemma 6.3.4.** *For $k$ a field of characteristic $p > 0$, $\mathrm{cor} : H^*(G, k) \to H^*(P \times G, k)$ vanishes.*

**Proof.** We have $\mathrm{cor} \circ \mathrm{res} = p\,\mathrm{Id} = 0$. However, since $P$ is a direct factor of $P \times G$, it follows that res is onto $H^*(G, k)$. Indeed, let $\iota : G \to P \times G$ denote the map identifying $G$ with $1 \times G$, and choose a splitting $\pi : P \times G \to G$ such that $\pi \circ \iota = \mathrm{Id}$. Then, by functorality, $\mathrm{res} \circ \pi^* = \iota^* \circ \pi^* = \mathrm{Id}$. □

**Corollary 6.3.5.** *Let $E$ be an elementary abelian $p$-group, $F$ a subgroup, and $k$ a field of characteristic $p$. Then $N_{F \to E} : H^{ev}(F, k) \to H^{ev}(E, k)$ is a ring homomorphism. (For $p = 2$, it is not necessary to restrict to even degrees.)*

**Proof.** Use the transitivity of the norm. □

Note that $N_{F \to E}$ is an $\mathbf{F}_p$-algebra homomorphism, but is not in general a $k$-algebra homomorphism since, for $a \in k$, we have $N_{F \to E}(a\alpha) = a^{(E:F)} N_{F \to E}(\alpha)$.

## 6.4 Serre's theorem

Let $G$ be a finite $p$-group. As in Section 3.3 and Exercise 4.2.3, we may identify

$$H^2(G, \mathbf{Z}) \cong \mathrm{Hom}(G, \mathbf{Q}/\mathbf{Z}),$$

so we may think of the non-trivial elements of order $p$ in $H^2(G, \mathbf{Z})$ as homomorphisms $\beta : G \to \mathbf{Q}/\mathbf{Z}$ with image $(1/p)\mathbf{Z}/\mathbf{Z} \cong \mathbf{Z}/p\mathbf{Z}$. These are exactly the $\beta$ with $H = \text{Ker}\,\beta$ a maximal normal subgroup of $G$ of index $p$. In fact, the basic theory of $p$-groups (Hall 1959, Theorem 4.3.2) tells us that the following conditions are equivalent for a $p$-group: (i) $H$ is maximal normal; (ii) $H$ is maximal; (iii) $H$ is of index $p$. It follows that any maximal subgroup $H$ of a $p$-group $G$ is the kernel of a $\beta$ of order $p$. Moreover, two such $\beta$'s with the same kernel differ by multiplication by an integer $r$ where $(r, p) = 1$. We shall abuse notation slightly by writing $\beta_H$ for any $\beta$ with $\text{Ker}\,\beta = H$. Let $G_2$ denote the Frattini subgroup of $G$, i.e., the intersection of all maximal subgroups of $G$ (Hall 1959, Section 10.4). Since there is a homomorphism from $G$ to some $(\mathbf{Z}/p\mathbf{Z})^d$ with kernel $G_2$, it follows that $G/G_2$ is elementary abelian. Similarly, it is not hard to see that if $G/H$ is elementary abelian, then $H \geq G_2$. (You can always find a collection of $\beta$'s of order $p$ whose kernels intersect in $H$.) It follows easily from the above facts that

$$\inf : H^2(G/G_2, \mathbf{Z}) = \text{Hom}(G/G_2, \mathbf{Q}/\mathbf{Z}) \to H^2(G, \mathbf{Z}) = \text{Hom}(G, \mathbf{Q}/\mathbf{Z})$$

is a monomorphism onto the subgroup of all $\beta$ with $p\beta = 0$.

There is another useful description of the Frattini subgroup. Since $G/G_2$ is abelian, $G_2 \geq [G, G]$, the subgroup of $G$ generated by commutators. Since $G/G_2$ has exponent $p$, it follows that $G_2 \geq G^p$, the subgroup of $G$ generated by $p$th powers. On the other hand, it is clear that $G/[G,G]G^p$ is elementary abelian so by the above reasoning $[G, G]G^p \geq G_2$. Hence, $G_2 = [G, G]G^p$.

**Theorem 6.4.1** (Serre). *Let $G$ be a finite $p$-group. $G$ is not elementary abelian if and only if there exist maximal subgroups $H_1, H_2, \ldots, H_k$ such that*

$$\beta_{H_1}\beta_{H_2}\ldots\beta_{H_k} = 0$$

*in $H^*(G, \mathbf{Z})$.*

**Proof.** If $G$ is elementary abelian, then by the results in Section 3.5 the subring of $H^*(G, \mathbf{Z})$ generated by $H^2(G, \mathbf{Z}) \cong \hat{G}$ is in positive degrees a symmetric algebra over $\mathbf{F}_p$. Hence, no product of non-trivial elements in $H^2(G, \mathbf{Z})$ can vanish.

Conversely, suppose $G$ is not elementary abelian, i.e., $G_2 \neq \{1\}$. Let $G_3 = [G, G_2]G_2{}^p$. $G_2/G_3$ is elementary abelian and central in $G/G_3$ by definition. It is also non-trivial. For, in general, if $G$ is a $p$-group, and $N$ is a non-trivial normal subgroup, then $N/[G, N]$ is non-trivial (Huppert 1967, Chapter III, Theorem 2.6), so its maximal elementary abelian factor group $N/N^p[G, N]$ is also non-trivial. Hence, we may choose a subgroup $M \geq G_3$ which is of index $p$ in $G_2$ and normal in $G$. Thus,

$$\mathbf{Z}/p\mathbf{Z} \cong G_2/M \to G/M \to G/G_2$$

is a central extension with kernel cyclic of order $p$, and *it does not split* since $G/M$ is not elementary abelian. We claim that it is enough to prove the theorem for $G/M$, i.e., we may assume $G_2$ is cyclic of order $p$. For,

$$\inf : H^2(G/M, \mathbf{Z}) = \operatorname{Hom}(G/M, \mathbf{Q}/\mathbf{Z}) \to H^2(G, \mathbf{Z}) = \operatorname{Hom}(G, \mathbf{Q}/\mathbf{Z})$$

is a monomorphism, so if there are non-trivial elements $\overline{\beta}_i \in H^2(G/M, \mathbf{Z})$ of order $p$ with product zero in $H^*(G/M, \mathbf{Z})$, their inflations $\beta_i \in H^2(G, \mathbf{Z})$ are non-trivial and have product 0.

Assume now that $G_2$ is cyclic of order $p$. We shall show that $G$ contains a subgroup $K$ which is not elementary abelian, with $K_2 = G_2$, and for which the theorem is true. This suffices by the following argument. Choose non-trivial elements $\beta_1, \ldots, \beta_r$ of order $p$ in $H^2(K, \mathbf{Z})$ with product 0. Since $K_2 = G_2$, we have $\beta_i = \inf(\overline{\beta}_i)$ for some $\overline{\beta}_i \in H^2(K/G_2, \mathbf{Z})$. Write $\overline{G} = G/G_2$ and $\overline{K} = K/G_2$. Since $(G : K) = (\overline{G} : \overline{K})$, it follows from Theorem 6.1.1 (N2), (N5) that

$$\begin{aligned}
0 = N_{K \to G}(0) &= N_{K \to G}(\prod \beta_i) \\
&= N_{K \to G}(\inf(\prod \overline{\beta}_i)) \\
&= \inf(N_{\overline{K} \to \overline{G}}(\prod \overline{\beta}_i)) \\
&= \inf(\prod N_{\overline{K} \to \overline{G}}(\overline{\beta}_i)).
\end{aligned}$$

However, by Corollary 6.3.2, $N_{\overline{K} \to \overline{G}}(\overline{\beta}_i)$ is a product of non-trivial elements of $H^2(G/G_2, \mathbf{Z})$ (necessarily of order $p$). Substituting in the above equation and inflating to $G$, we have the desired result.

We now show the existence of the desired subgroup $K$. If $G$ contains a cyclic subgroup $K$ of order $p^2$, we are done. For, in that case, $\{1\} < K_2 = K^p \leq G^p \leq G_2 \cong \mathbf{Z}/p\mathbf{Z}$ and so $K_2 = G_2$. Also, $H^2(K, \mathbf{Z}) = \operatorname{Hom}(K, \mathbf{Q}/\mathbf{Z})$ is generated by an element $\chi$ of order $p^2$, so $\beta = p\chi$ is of order $p$ and satisfies $\beta^2 = p^2 \chi^2 = 0$.

Hence, we may assume that every non-trivial element of $G$ has order $p$, and, since $G$ is not cyclic or elementary abelian, that $G/G_2$ has rank at least 2. Consider subgroups $G \geq K > G_2$ such that $\overline{K} = K/G_2$ has rank 2. At least one such $K$ is not abelian or it would follow that every pair of elements of $G$ commutes, and, since every element has order $p$, $G$ would be elementary abelian. Then, at least one commutator $[x, y] \neq 1$ with $x, y \in K$, and it follows easily that $K_2 = G_2$. We shall verify the theorem for such $K$.

Let $H > K_2$ be a maximal subgroup of $K$. $H$ is necessarily abelian (since it is of order $p^2$), and it is not cyclic, so $H$ is elementary abelian of rank 2. Let $H^2(H, \mathbf{Z}) = \mathbf{Z}\eta \oplus \mathbf{Z}\zeta$. Let $xH$ generate the cyclic group $K/H$. We may suppose that conjugation by $x$ is a non-trivial automorphism of

$H$ or else $K$ would be abelian. By standard linear algebra, we know that, up to a change of basis in $H$, there is only one way an element of order $p$ can act non-trivially on $\hat{H} = H^2(H, \mathbf{Z})$, so we may suppose

$$x^*(\eta) = \eta$$
$$x^*(\zeta) = \eta + \zeta.$$

Note that if $yK_2$ generates $H/K_2$, we must have $x^{-1}yx = yz$ where $z$ generates $K_2$. (Otherwise, it would follow that $x$ acts trivially on $H$.) Hence,

$$\eta(z) = \eta(x^{-1}yxy^{-1}) = x^*(\eta)(y) - \eta(y) = 0$$

so $\eta(G_2) = 0$, and $\eta = \inf \bar{\eta}$ where $\bar{\eta}$ generates $H^2(H/G_2, \mathbf{Z})$. It follows as above that $N_{H\to K}(\eta)$ is a product of inflations of non-trivial elements of $H^2(K/K_2, \mathbf{Z})$, i.e., it is a product of non-trivial elements of $H^2(K, \mathbf{Z})$ of order $p$.

Choose $\alpha \in H^2(K, \mathbf{Z})$ with kernel $H$. To complete the proof, it suffices to show that

$$\alpha N_{H\to K}(\eta) = 0.$$

To see this, argue as follows. Since $x$ acts trivially on $H^*(K, \mathbf{Z})$, we have

$$N(\zeta) = x^* N(\zeta) = N(x^*\zeta) = N(\eta + \zeta)$$
$$= N(\eta) + \mathrm{cor}\,\mu + N(\zeta)$$

where $\mu \in H^*(H, \mathbf{Z})$. It follows that $N(\eta) = -\mathrm{cor}\,\mu$ so

$$\alpha N(\eta) = -\alpha \mathrm{cor}(\mu) = -\mathrm{cor}(\mathrm{res}(\alpha)\,\mu) = -\mathrm{cor}(0\,\mu) = 0. \quad \square$$

The above proof relies in part on work of Okuyama and Sasake (1990). The argument clearly depends on some understanding of the integral cohomology rings of non-elementary abelian groups of order $p^3$. For $p = 2$, such a group is either $\mathbf{Z}/2\mathbf{Z} \times \mathbf{Z}/4\mathbf{Z}$ or the dihedral or quaternion group of order 8. These cohomology rings are known. (See Cartan and Eilenberg (1956, Section XII.7) for quaternion groups and Evens (1965, Section 5) for the dihedral group.) In each such case there is a cyclic normal subgroup of order 4 which in the above argument is the simpler case. For $p$ odd, one of the non-elementary abelian groups of order $p^3$ has exponent $p$, and it is basically for that case that we need the last argument. (See Lewis (1968) for a complete description of the integral cohomology ring of that group.)

Serre originally stated the theorem as follows.

**Corollary 6.4.2** (Serre). *Let $G$ be a projective limit of finite $p$-groups. $G$ is not elementary abelian if and only if there exist non-trivial elements $\alpha_1, \alpha_2, \ldots, \alpha_r \in H^1(G, \mathbf{F}_p) = \text{Hom}(G, \mathbf{F}_p)$ such that for $p = 2$*

$$\alpha_1 \alpha_2 \ldots \alpha_r = 0$$

*or for $p$ odd the product of the Bocksteins*

$$\delta\alpha_1 \delta\alpha_2 \ldots \delta\alpha_r = 0.$$

**Outline of a Proof.** The reduction from projective limits of $p$-groups to $p$-groups poses no difficulties. Also, one can derive the original form from the version above without too much difficulty. In particular, the elements of order $p$ in $H^2(G, \mathbf{Z})$ may be viewed as integral Bocksteins of elements $\alpha \in H^1(G, \mathbf{F}_p)$ and their reductions modulo $p$ are the Bocksteins $\delta\alpha$. The form of the theorem for $p = 2$ follows if one realizes that in that case $\delta\alpha = \alpha^2$ for $\alpha \in H^1(G, \mathbf{F}_2)$. Alternately, one may just reprove the theorem as above for $p = 2$ taking coefficients in $\mathbf{F}_2$ and realizing that everything commutes so the norm arguments may be used directly for elements of degree 1. □

Serre's (1965a) proof used an elegant geometric argument in which the polynomial subring of $H^*(G/G_2, \mathbf{F}_p)$ generated by $\delta H^1(G/G_2, \mathbf{F}_p)$ is considered as the coordinate ring of affine $n$-space. He took as his starting point a relation among the polynomial generators derived using Steenrod operations. Serre (1987) later presented an independent proof which does not use algebraic geometry. The theorem has also been proved by Kroll (1985).

All the more recent proofs of Serre's Theorem allow one to determine an upper bound for the minimum number of terms in a vanishing product. We leave it to the reader to determine the number implicit in our proof.

**Exercise 6.4.1.** Carefully analyse the proof of Theorem 6.4.1 to find an upper bound for the degree in which the relation occurs. Distinguish the case $p = 2$ and the case $p$ is odd.

# 7
# Spectral sequences

## 7.1 The spectral sequence of a double complex

Spectral sequences play an important role in group cohomology because they provide a means of reducing cohomology in a complex situation to the cohomology of constituents. The most important spectral sequence for us will be the Lyndon–Hochschild–Serre (LHS) spectral sequence which relates the cohomology of a group to that of a normal subgroup and that of the factor group. We have seen two special cases of this situation. The Künneth Theorem relates the cohomology of $K \times H$ to that of $K$ and $H$; in particular if $k$ is a field, then

$$H^*(K \times H, k) \cong H^*(K, k) \otimes H^*(H, k).$$

Similarly, if $k$ is a field, Nakaoka's Theorem (Theorem 5.3.1) tells us that the cohomology of the semi-direct product $S \int H = S \rtimes H^{|X|}$ is given by

$$H^*(S \rtimes H^{|X|}, k) \cong H^*(S, H^*(H^{|X|}, k)) \cong H^*(S, H^*(H, k)^{\otimes X}).$$

In general, if $G$ is a group, $H$ a normal subgroup, and $M$ a $kG$-module, there is a spectral sequence approximating $H^*(G, M)$ which starts with

$$H^*(G/H, H^*(H, M)).$$

The reader should be familiar with the notion of a spectral sequence and how such a sequence is associated with the cohomology of a filtered complex. There are many good expositions of this subject. (See for example Mac Lane (1963, Chapter XI) and Hilton and Stammbach (1971, Chapter VIII).) Almost every spectral sequence used in this monograph will arise from a double complex, so we shall review the theory in that context. The reader who is uneasy with spectral sequences could start with the review and then check the references for details.

Let $A = \{A^{p,q}, d', d''\}$ denote a double (cochain) complex. By this we mean that $d'$ is a map of bidegree $(1,0)$, $d''$ is a map of bidegree $(0,1)$, $d'^2 = d''^2 = 0$, and $d'd'' + d''d' = 0$. We let $A^n = \bigoplus_{p+q=n} A^{p,q}$, and $d = d' + d''$ so that $\{A^n, d\}$ is a single complex. When we talk about the cohomology of $A$, we mean the cohomology of this associated single complex. In essentially everything we shall do, we can assume the non-zero components of $A$ are restricted to one of the quadrants of the $p,q$-plane — usually the first quadrant. *In what follows assume unless otherwise specified that $A^{p,q}$ vanishes outside the first quadrant.* Think of $d'$ going horizontally, $d''$ going vertically, and $A^n$ obtained by adding up along the line $p+q=n$.

The cohomology of $A$ may be approximated as follows. First, calculate the cohomology $H''(A)$ with respect to the vertical differential. Since $d'd'' = -d''d'$, the horizontal differential $d'$ induces a differential on $H''(A)$ and we may then compute its cohomology $H'H''(A)$. (Of course, we can also do this in the other order $H''H'(A)$ but the results will not generally be the same.) If the cohomology $H^*(A)$ of the associated single complex is $H'H''(A)$, there is not much else to do, but that is seldom the case. More commonly, we need to extend this process so as to approximate $H^*(A)$. To see how to do this, first consider in detail how $H'H''(A)$ is computed. Start in the $(p,q)$ position. Let $a^{p,q}$ be a *vertical* cocycle, i.e. $d''a^{p,q}=0$. It defines a class in $H''(A)$ modulo the image under $d''$ of elements in the $(p,q-1)$ position. For $a^{p,q}$ to represent a (*horizontal*) cocycle in $H''(A)$ under $d'$, it must be true that $d'a^{p,q}$ (which sits at coordinates $(p+1,q)$) is the image under $d''$ of an element $a^{p+1,q-1}$ with coordinates $(p+1,q-1)$. Thus

$$d(a^{p,q} - a^{p+1,q-1}) = -d'a^{p+1,q-1} \in A^{p+2,q-1},$$

so the expression $a^{p,q} - a^{p+1,q-1}$ is a cocycle modulo everything two steps to the right of the $(p,q)$ position. Similarly, it represents a coboundary in $H''(A)$ under $d'$ if there are elements $b^{p-1,q}$ and $b^{p,q-1}$ such that

$$d''b^{p-1,q} = 0$$
$$d'b^{p-1,q} = d''b^{p,q-1} + a^{p,q}$$

i.e.

$$d(b^{p-1,q} - b^{p,q-1}) \equiv a^{p,q}$$

modulo everything two steps to the right of coordinates $(p-1,q)$ or one step to the right of coordinates $(p,q)$.

It is now clear how to extend this approximation procedure. (We shall use the general approach of Mac Lane (1963) but with slightly different notation.) Filter the complex $A$ by letting $F^pA$ be the double subcomplex whose components to the left of the $p$th column are zero, i.e.

$$F^pA^n = \bigoplus_{p'\geq p} A^{p',n-p'}.$$

Note that $F^0 A^n = A^n$, and $F^p A^n = 0$ for $p > n$ since that takes us out of the first quadrant. Let $C_r^{p,q}$ be the set of those elements in $F^p A^{p+q}$ for which the *total* differential belongs to $F^{p+r} A^{p+q+1}$. Each such element should be visualized as a sum of components along the line $p + q = n$, starting at the $(p, q)$ position, going down and to the right, and such that the vertical and horizontal differentials cancel *within the band* $p \leq p' < p + r$. Note that the total differential of such an element lies in $F^{p+r} A^{n+1}$, i.e. it starts at coordinates $(p+r, q-r+1)$. Define

$$E_r^{p,q} = \frac{C_r^{p,q} + F^{p+1} A^{p+q}}{d(C_{r-1}^{p-r+1, q+r-2}) + F^{p+1} A^{p+q}}.$$

From the above discussion, one is not surprised to learn that $d$ induces morphisms

$$d_r^{p,q} : E_r^{p,q} \to E_r^{p+r, q-r+1}$$

satisfying $d_r^2 = 0$. Also, if we compute the cohomology of the resulting complex, it turns out that

$$H(E_r, d_r) = E_{r+1},$$

i.e. $E_{r+1}^{p,q} = \mathrm{Ker}\, d_r^{p,q} / \mathrm{Im}\, d_r^{p-r, q+r-1}$. The reader should carefully study the diagram implicit in the above discussion to understand the structure of a representative $a$ of an element in $E_r^{p,q}$. Note that each such representative $a$ defines an element in a *subquotient* of $A^{p,q}$ at its upper left $(p,q)$, but its extended structure to the right of $(p,q)$ is important in calculating $d_r$. In particular, $da \in F^{p+1} A$ represents $d_r$ of the element represented by $a$.

For each fixed position $(p,q)$ the differentials $d_r^{p,q}$ which start there and the differentials $d_r^{p-r, q+r-1}$ which end there must vanish for $r$ sufficiently large since they either start or end outside the first quadrant. It follows that each $E_r^{p,q}$ eventually stabilizes at a common value which we denote $E_\infty^{p,q}$. Note, however, that the value of $r$ at which stabilization occurs usually depends on $p$ and $q$ and there may not be any single $r$ such that $E_r = E_\infty$ for all $(p,q)$. Suppose now that $a \in A^n$ is a cocycle which starts at $A^{p,q}$ where $p + q = n$; i.e. $a \in F^p A^n$ but $a \notin F^{p+1} A^n$ and $da = 0$. Clearly, $a$ determines an element of $E_r^{p,q}$ for every $r \geq 1$, and $d_r$ is zero on that element, so $a$ determines an element of $E_\infty^{p,q}$. In other words, we have a morphism

$$F^p H^{p+q}(A) = \mathrm{Im}\{H^{p+q}(F^p A) \to H^{p+q}(A)\} \to E_\infty^{p,q}$$

which is an epimorphism. Moreover, it is not hard to see that its kernel is $F^{p+1} H^{p+q}(A)$. Thus, $H^*(A)$ is filtered by $F^p H^*(A)$ and

$$F^p H^n(A) / F^{p+1} H^n(A) \cong E_\infty^{p,q}(A).$$

Thus, the spectral sequence only gives us information about $H^*(A)$ modulo this filtration.

The above constructions and definitions are completely natural with respect to whatever you might reasonably want to do with a double complex. In particular, suppose we have a multiplicative structure. For example, $A$ might be a doubly graded algebra such that $d'$ and $d''$ are both derivations with respect to the *total* degree. (More generally, we might have three double complexes $A$, $B$, and $C$, and a pairing $A \otimes B \to C$ with appropriate properties for the differentials.) Then, it is easy to see that the multiplicative structure is inherited at each stage of the spectral sequence, the $d_r$ are derivations, and all the relevant morphisms are consistent with the multiplicative structures. Notice in particular that if $A$ is a doubly graded algebra, then the spectral sequence gives us information about the associated graded ring $Gr(H^*(A))$.

## 7.2 The LHS spectral sequence of a group extension

Let $G$ be a group, $H$ a normal subgroup, and $M$ a $kG$-module. We shall construct a spectral sequence with $E_2$ term $H^*(G/H, H^*(H, M))$ and with $E_\infty$ term $Gr(H^*(G, M))$ (for a suitable filtration of $H^*(G, M)$.) This spectral sequence was anticipated by Lyndon (1948), and it was worked out in detail by Hochschild and Serre (1953). Their constructions were based on the bar resolution so were quite involved and left some issues unresolved. (See Evens (1975, Section 4) and Beyl (1981).) Our approach is based on an appropriate double complex and is relatively straightforward.

We form a double complex as follows. Let $X \to k$ be a $kG$-projective resolution and let $Y \to k$ be a $k(G/H)$-projective resolution. Then $X \to k$ is also a $kH$-projective resolution. Recall from Section 1.1 that $G$ acts on $\operatorname{Hom}_H(X, M)$ by $(gf)(x) = g(f(g^{-1}x))$, and since $H$ then acts trivially, we may view $\operatorname{Hom}_H(X, M)$ as a $k(G/H)$-complex. Form the double complex

$$A = \operatorname{Hom}_{G/H}(Y, \operatorname{Hom}_H(X, M)).$$

As in Section 1.2,

$$d' = \operatorname{Hom}_{G/H}(d_Y, \operatorname{Hom}_H(\operatorname{Id}, \operatorname{Id}))$$
$$d'' = \operatorname{Hom}_{G/H}(\operatorname{Id}, \operatorname{Hom}_H(d_X, \operatorname{Id})),$$

but by the Cartan–Eilenberg sign convention there is an implied *sign* on the right hand side of the definition of $d''$. (It is $(-1)^p$ where $p$ denotes the degree of the $Y$ argument.) As usual in a double complex, $d'^2 = d''^2 = 0$ and $d'd'' + d''d' = 0$.

As in the previous section, the double complex $A$ yields a spectral sequence with $E_2 = H'H''(A)$ and $E_\infty = Gr(H^*(A))$, the associated graded object arising from the filtration $F^p H^*(A)$. On the other hand, we can reverse the roles of the horizontal index $p$ and the vertical index $q$ to obtain a second spectral sequence starting with $H''H'(A)$. The filtration in this case is obtained by adding everything *on or above* a given horizontal line at $q$. We call these spectral sequences the first and second spectral sequences of the complex. Both converge to the cohomology of the associated single complex $H^*(A)$.

We shall calculate the $E_2$ terms which start these spectral sequences. In particular, we shall show that the first spectral sequence starts with

$$H^*(G/H, H^*(H, M))$$

where $H^*(H, M)$ is viewed as a $G$-module and hence a $G/H$-module as above. Moreover, we shall show that the second spectral sequence 'collapses' in the sense that most of the $E_2$-terms are trivial allowing us to calculate $H^*(A)$; it turns out to be $H^*(G, M)$. This is a common strategy for spectral sequence calculations: use one of two spectral sequences for the same object to identify it as the cohomology of something interesting and show the other starts with an interesting $E_2$ term.

*The first spectral sequence $H'H''(A)$.*
We have

$$H''(\mathrm{Hom}_{G/H}(Y, \mathrm{Hom}_H(X, M))) = \mathrm{Hom}_{G/H}(Y, H^*(\mathrm{Hom}_H(X, M)))$$

since $Y$ is $k(G/H)$-projective and $\mathrm{Hom}_{G/H}(Y, -)$ is an exact functor. Thus,

$$E_2 = H'H''(A) = H^*(\mathrm{Hom}_{G/H}(Y, H^*(H, M))) = H^*(G/H, H^*(H, M))$$

as claimed above.

*The second spectral sequence $H''H'(A)$.*
We have

$$H'(\mathrm{Hom}_{G/H}(Y, \mathrm{Hom}_H(X, M))) = H^*(G/H, \mathrm{Hom}_H(X, M)).$$

**Lemma 7.2.1.** $H^p(G/H, \mathrm{Hom}_H(X, M)) = 0$ for $p > 0$.

**Proof.** Since each $X_q$ is $kG$-projective, hence a direct summand of a free $kG$-module, it suffices to prove the lemma for $X = kG$.

Let $\tilde{M}$ denote the underlying $k$-module for $M$ but with trivial $G$-action. We shall define an isomorphism of $kG$-modules

$$\mathrm{Hom}_H(kG, M) \cong \mathrm{Hom}_H(kG, \tilde{M})$$

where $G$ acts on the left hand side by $(gf)(x) = gf(g^{-1}x)$, but on the right hand side as on the *coinduced module*: $(gf')(x) = f'(xg)$.

For $f \in \text{Hom}(kG, M)$, define $f' \in \text{Hom}(kG, \tilde{M})$ by $f'(x) = xf(x^{-1})$. Clearly, $(f')' = f$. In addition, if $f \in \text{Hom}_H(kG, M)$, then

$$f'(xh) = xhf(h^{-1}x^{-1}) = xf(x^{-1}) = f'(x).$$

Hence, $f'(hx) = f'(xx^{-1}hx) = f'(x)$ because $H$ is normal in $G$, and $f' \in \text{Hom}_H(kG, \tilde{M})$. Conversely, if $f' \in \text{Hom}_H(kG, \tilde{M})$, it is easy to see that $f = f'' \in \text{Hom}_H(kG, M)$. Thus, we have an isomorphism $f \leftrightarrow f'$ of $k$-modules. To check that it is a $G$-isomorphism, note that

$$(gf)'(x) = x(gf)(x^{-1}) = xgf(g^{-1}x^{-1}) = f'(xg) = (gf')(x).$$

On the other hand, since $H$ acts trivially on $\tilde{M}$, we have

$$\text{Hom}_H(kG, \tilde{M}) \cong \text{Hom}(k(G/H), \tilde{M})$$

so, by the Eckmann–Shapiro Lemma (Proposition 4.1.3),

$$H^p(G/H, \text{Hom}_H(kG, M)) \cong H^p(G/H, \text{Hom}(k(G/H), \tilde{M}))$$
$$\cong H^p(1, \tilde{M}) = 0$$

for $p > 0$. □

It follows from the lemma that $H'(A)$ is concentrated on the line $p = 0$, and on that line it is

$$H^0(G/H, \text{Hom}_H(X, M)) = \text{Hom}_H(X, M)^{G/H} = \text{Hom}_G(X, M).$$

Hence, $H''H'(A) = H^*(\text{Hom}_G(X, M)) = H^*(G, M)$.

Since the $E_2$ term of the second spectral sequence is concentrated on one line, it follows that $E_r = E_\infty$ for $r \geq 2$, and $E_\infty$ is also concentrated on the line $p = 0$. Hence the filtration of $H^n(A)$ has only one non-trivial quotient and we conclude that $H^n(A) = H^n(G, M)$ for each $n$. (Warning: Because of the switch between horizontal and vertical for the second spectral sequence, we actually have $F^0 H^n = F^n H^n$ and $E_\infty^{n,0} = F^n H^n$, with the other $E_\infty$ terms being 0.)

**Exercise 7.2.1.** Let $G$ be a group, let $M$ and $N$ be $\mathbf{Z}G$-modules, and let $X \to \mathbf{Z}$ and $P \to M$ be $\mathbf{Z}G$-projective resolutions.

(a) By considering the double complex $\text{Hom}_{\mathbf{Z}G}(X, \text{Hom}_{\mathbf{Z}}(P, N))$, construct a spectral sequence

$$H^*(G, \text{Ext}^*_{\mathbf{Z}}(M, N)) \Longrightarrow \text{Ext}^*_{\mathbf{Z}G}(M, N).$$

(Hint: $P \to M$ is also a **Z**-free resolution.)

(b) Using the fact that $\operatorname{Ext}^s_{\mathbf{Z}}(M, N) = 0$ for $s > 1$, show that the spectral sequence amounts to a long exact sequence

$$\cdots \to H^n(G, \operatorname{Hom}(M, N)) \to \operatorname{Ext}^n_{\mathbf{Z}G}(M, N) \to$$
$$H^{n-1}(G, \operatorname{Ext}(M, N)) \to H^{n+1}(G, \operatorname{Hom}(M, N)) \to \cdots$$

*Functorial dependence.*

Let $H$ be a normal subgroup of $G$, and let $H'$ be a normal subgroup of $G'$. Let $\phi : G' \to G$ be a group homomorphism such that $\phi(H') \subseteq H$. Then, $\phi$ induces a homomorphism $\overline{\phi} : G'/H' \to G/H$. Finally, let $M$ be a $kG$-module, $M'$ a $kG'$-module, and $f : M \to M'$ a homomorphism such that $g'f(m) = f(\phi(g')m)$ for $g' \in G', m \in M$.

Choose resolutions $X \to k$ and $Y \to k$ for $G$ and $G/H$ as above. Similarly, choose resolutions $X' \to k$ and $Y' \to k$ for $G'$ and $G'/H'$. Then there are chain maps $X' \to X$ and $Y' \to Y$ consistent with the augmentations and the above group homomorphisms. Moreover, these maps are unique up to chain homotopy. All these maps together induce a morphism of double complexes

$$\operatorname{Hom}_{G/H}(Y, \operatorname{Hom}_H(X, M)) \to \operatorname{Hom}_{G'/H'}(Y', \operatorname{Hom}_{H'}(X', M'))$$

and different choices of the chain maps yield *bihomotopic* morphisms. (See Cartan and Eilenberg (1956, IV.4) for a discussion of 'bihomotopy'.) These maps in turn induce maps of the objects constructed in the two spectral sequences. Thus, they induce well defined maps for all terms of both spectral sequences for $r \geq 2$. (For $E_1$ there is still the possibility of variation by a chain homotopy.) Moreover, it is not hard to see that the induced maps at the levels at which we have identified the terms are exactly the maps we would expect. In particular, in the second spectral sequence, the map induced by the isomorphisms

$$H^*(G, M) \cong H^*(A)$$
$$H^*(G', M') \cong H^*(A')$$

is just the functorial map $(\phi, f)^* : H^*(G, M) \to H^*(G', M')$. Similarly, the map induced at the $E_2$ level of the first spectral sequence is the functorial map

$$(\overline{\phi}, (\phi', f)^*)^* : H^*(G/H, H^*(H, M)) \to H^*(G'/H', H^*(H', M')).$$

Finally, all the relevant maps and identifications in the discussion of the spectral sequences are completely natural with respect to change of groups and modules, and all the obvious diagrams commute.

## Edge homomorphisms

Consider the first spectral sequence of a double complex $A$. There are two natural maps associated with the edges. For the horizontal edge, $H''(A^{*,0}) = \operatorname{Ker} d^{*,0}$ is a subcomplex of $A$ with differential identifiable as $d'$.

Thus, there is a morphism

$$E_2^{*,0} = H'(H''(A^{*,0})) \to H^*(A).$$

On the other hand, $F^0 H^*(A)/F^1 H^*(A) = E_\infty^{0,*}$ is a *subgroup* of $E_2^{0,*}$ since no non-trivial $d_r$ end in the vertical edge. It follows that there is a natural map

$$H^*(A) = F^0 H^*(A) \to E_2^{0,*}.$$

These two maps are called the *edge homomorphisms*. For the other spectral sequence, the roles of horizontal and vertical are just reversed.

**Proposition 7.2.2.** *Let $H$ be a normal subgroup of $G$ and let $M$ be a $kG$-module. The horizontal edge homomorphism in the first spectral sequence*

$$H^*(G/H, M^H) = E_2^{*,0} \to H^*(A) = H^*(G, M)$$

*is* $\operatorname{inf}_{G/H \to G}$. *The vertical edge homomorphism*

$$H^*(G, M) = H^*(A) \to E_2^{0,*} = H^*(H, M)^{G/H} \subseteq H^*(H, M)$$

*is* $\operatorname{res}_{G \to H}$.

**Proof.** For the horizontal edge homomorphism, consider the relation between the triples $(G, H, M)$ and $(G/H, 1, M^H)$ and look at the induced maps of spectral sequences. For the vertical edge homomorphism consider the induced map of spectral sequences for the triples $(H, H, M)$ and $(G, H, M)$. The details are left to the reader. Note, however, that the proofs require a bit more than 'general nonsense'. In particular, you have to show that for a triple of the form $(K, 1, N)$, the horizontal edge homomorphism is the identity. That requires some explicit discussion at the level of the resolutions $Y$ and $X$ which in that case are both $kK$-resolutions. □

**Exercise 7.2.2.** Let $K$ be a group and $N$ a $K$-module. Show explicitly that the horizontal edge homomorphism for the triple $(K, 1, N)$

$$H^*(K/1, N^1) \to H^*(K, N)$$

is the identity.

**Exercise 7.2.3.** Let $G$ be a finite group and $N$ a normal Hall subgroup, i.e. assume $|N|$ and $|G/N|$ are relatively prime. Let $M$ be any $G$-module. Show that, for each $n > 0$, the sequence provided by the edge homomorphisms (inflation and restriction)

$$0 \to H^n(G/N, M^N) \to H^n(G, M) \to H^n(N, M)^{G/N} \to 0$$

is exact and splits. (Note that $G \cong (G/N) \ltimes N$ by the Schur–Zassenhaus Theorem.)

**Corollary 7.2.3.** *Let $H$ be a normal subgroup of $G$ and let $M$ be a $kG$-module. Then there is an exact sequence*

$$0 \to H^1(G/H, M^H) \xrightarrow{\text{inf}} H^1(G, M) \xrightarrow{\text{res}} H^1(H, M)^{G/H}$$
$$\xrightarrow{d_2^{0,1}} H^2(G/H, M^H) \xrightarrow{\text{inf}} H^2(G, M).$$

The fourth map $d_2^{0,1}$ is usually called *transgression*.

**Proof.** We have

$$0 \to H^1(G/H, M^H) \cong E_2^{1,0} = E_\infty^{1,0} = F^1 H^1(G, M) \to H^1(G, M)$$
$$\to H^1(G, M)/F^1 H^1(G, M) = E_\infty^{0,1} = \operatorname{Ker} d_2^{0,1} \to 0$$

and

$$0 \to \operatorname{Ker} d_2^{0,1} \to E_2^{0,1} \cong H^1(H, M)^{G/H}$$
$$\to E_2^{2,0} \cong H^2(G/H, M^H)$$
$$\to E_\infty^{2,0} = F^2 H^2(G, M) \to 0. \quad \square$$

The five-term exact sequence in the corollary is often called *the fundamental exact sequence*. It is a basic tool in many important applications of group cohomology. For example, part of it is used in the proof of the Tate–Nakayama Theorem (Atiyah and Wall 1967, Theorem 10) which plays a central role in the cohomological formulation of class field theory. Because of its importance, it was discussed extensively in the early literature where it was proved without benefit of spectral sequences. It may even be found in a rudimentary but disguised form in the work of Schur (1904, Section 2, Theorem II).

We conclude this section with some applications of the fundamental sequence.

**Theorem 7.2.4.** *Let $k$ be a field of characteristic $p$ and let $G$ and $H$ be $p$-groups. Suppose $\phi : H \to G$ is a homomorphism of groups such that $\phi^* : H^1(G,k) \to H^1(H,k)$ is an isomorphism and $\phi^* : H^2(G,k) \to H^2(H,k)$ is a monomorphism. Then $\phi$ is an isomorphism.*

**Proof.** Form a central series for $G$ as follows. As in Section 6.4, let $G_2 = [G,G]G^p$ be the Frattini subgroup of $G$, i.e. the minimal normal subgroup of $G$ such that $G/G_2$ is elementary abelian. Assume $G_n$ has been defined, and define $G_{n+1}$ to be the smallest normal subgroup of $G$ which is contained in $G_n$ and such that $G_n/G_{n+1}$ is elementary abelian and is contained in the centre of $G/G_{n+1}$. ($G_{n+1}$ is generated by all commutators $[g,h]$ where $g \in G, h \in G_n$ and all $h^p$ with $h \in G_n$. As above, this is denoted symbolically by $G_{n+1} = [G,G_n]G_n{}^p$.) It is not hard to see from basic theorems about $p$-groups that this series terminates at the trivial group $\{1\}$ for some $n$. To prove the theorem, we shall show inductively that $\phi$ induces an isomorphism $H/H_n \cong G/G_n$ for every $n$.

For $n = 2$,

$$H^1(G,k) = \operatorname{Hom}(G,k) \cong \operatorname{Hom}(G/G_2, k) = \operatorname{Hom}_{\mathbf{F}_p}(G/G_2, k)$$

(since $G/G_2$ is an $\mathbf{F}_p$-vector space). However, $\operatorname{Hom}_{\mathbf{F}_p}(X,k) = 0$ if and only if $X = 0$, so by the exactness of the functor $\operatorname{Hom}_{\mathbf{F}_p}$, it follows that the kernel and cokernel of $H/H_2 \to G/G_2$ are trivial.

Consider the commutative diagram of group extensions

$$\begin{array}{ccccccccc} 1 & \to & G_n & \to & G & \to & G/G_n & \to & 1 \\ & & \uparrow & & \uparrow & & \uparrow & & \\ 1 & \to & H_n & \to & H & \to & H/H_n & \to & 1 \end{array}$$

Since the LHS spectral sequence is functorial, it follows that we have a commutative diagram relating the five-term exact sequences resulting from the two spectral sequences. In addition, we have

$$H^1(G_n, k)^G \cong \operatorname{Hom}(G_n, k)^G \cong \operatorname{Hom}(G_n/[G,G_n], k)$$
$$\cong \operatorname{Hom}(G_n/G_{n+1}, k) = H^1(G_n/G_{n+1}, k)$$

by the *construction* of $G_{n+1}$, and similarly for $H$. Hence, we have the following commutative diagram

$$\begin{array}{ccccccccc} H^1(G/G_n) & \to & H^1(G) & \to & H^1(G_n/G_{n+1}) & \to & H^2(G/G_n) & \to & H^2(G) \\ \downarrow & & \downarrow & & \downarrow & & \downarrow & & \downarrow \\ H^1(H/H_n) & \to & H^1(H) & \to & H^1(H_n/H_{n+1}) & \to & H^2(H/H_n) & \to & H^2(H) \end{array}$$

where we have omitted the coefficient module $k$. By induction, the first and fourth vertical arrows are isomorphisms. By hypothesis, the second is an isomorphism and the fifth is a monomorphism. Hence, by the five-lemma, the third arrow is an isomorphism. It follows that $H_n/H_{n+1} \cong G_n/G_{n+1}$ since both these groups are elementary abelian.

By considering the diagram

$$\begin{array}{ccccccccc} 1 & \longrightarrow & G_n/G_{n+1} & \longrightarrow & G/G_{n+1} & \longrightarrow & G/G_n & \longrightarrow & 1 \\ & & \uparrow & & \uparrow & & \uparrow & & \\ 1 & \longrightarrow & H_n/H_{n+1} & \longrightarrow & H/H_{n+1} & \longrightarrow & H/H_n & \longrightarrow & 1 \end{array}$$

it follows easily that $H/H_{n+1} \cong G/G_{n+1}$ as claimed. □

The above theorem is a modification of a theorem of Stallings (1965). Stallings considered the case $k = \mathbf{Z}$ and his theorem uses the lower central series of the group instead of the 'Frattini $p$-series' we considered.

**Corollary 7.2.5.** *If $G$ is a $p$-group and* $\inf : H^2(G/G_2, k) \to H^2(G, k)$ *is a monomorphism, then $G_2 = 1$, i.e. $G$ is elementary abelian.*

**Proof.** The assertion follows immediately from the theorem by taking $G$ to be $G/G_2$ and $H$ to be $G$, but it may also be proved directly quite easily using only a small part of the argument used to prove the theorem. □

Note that the corollary may be considered the starting point for the argument in the proof of Serre's Theorem.

A second application of the fundamental sequence concerns the relation between a finite group and its Sylow subgroups. Let $G$ be finite and suppose $p$ is a prime dividing $|G|$. Let $U$ be the subgroup of $G$ generated by all elements of order relatively prime to $p$. It is clear that $U$ is normal and $G/U$ is a $p$-group. Moreover, if $P$ is any $p$-Sylow subgroup of $G$, then $PU = G$, since $U$ contains every $q$-Sylow subgroup of $G$ for $q \neq p$ so $|PU| \geq |G|$. Hence, $G/U \cong P/P \cap U$. In finite group theory, one is interested in circumstances in which $P \cap U = \{1\}$, i.e. $G/U \cong P$. It is easy to see that this is equivalent to saying $G \cong P \ltimes U$ for some normal subgroup $U$. A finite group with this property is called $p$-*nilpotent*. Finite group theorists are interested in criteria for establishing $p$-nilpotence, and one such theorem is due to Tate (1963).

**Theorem 7.2.6** (Tate). *Let $G$ be a finite group and $P$ a $p$-Sylow subgroup for some prime $p$. If $G/G_2 \cong P/P_2$, then $G$ is $p$-nilpotent.*

**Proof.** First note that $H^1(G, \mathbf{F}_p) \cong \mathrm{Hom}(G/G_2, \mathbf{F}_p)$ and

$$\mathrm{res}_{G \to P} : H^n(G, \mathbf{F}_p) \to H^n(P, \mathbf{F}_p)$$

is a monomorphism in any case. Hence, the hypothesis is equivalent to restriction being an isomorphism in degree 1. Put $V = P \cap U$ and consider the diagram of extensions

$$\begin{array}{ccccc} U & \longrightarrow & G & \longrightarrow & G/U \\ \uparrow & & \uparrow & & \uparrow \cong \\ V & \longrightarrow & P & \longrightarrow & P/V \end{array}$$

This induces a commutative diagram of fundamental sequences (with all coefficients in $\mathbf{F}_p$)

$$\begin{array}{ccccccccc} H^1(G/U) & \to & H^1(G) & \to & H^1(U)^G & \to & H^2(G/U) & \to & H^2(G) \\ \cong \downarrow & & \cong \downarrow & & \downarrow & & \cong \downarrow & & \text{mono} \downarrow \\ H^1(P/V) & \to & H^1(P) & \to & H^1(V)^P & \to & H^2(P/V) & \to & H^2(P) \end{array}$$

which by the five-lemma tells us that $H^1(U)^G \to H^1(V)^P$ is an isomorphism. However, since $U$ is generated by elements prime to $p$, it is clear that $H^1(U) = \text{Hom}(U, \mathbf{F}_p) = \{0\}$. Hence, $H^1(V)^P = \{0\}$, and since $P$ is a $p$-group, that implies that $H^1(V) = \{0\}$, whence $V = \{1\}$ as required. $\square$

## 7.3 Multiplicative structure in the spectral sequence

Let $G$ be a group, $H$ a normal subgroup, and $M$ and $N$ $kG$-modules. Choose a $kG$-projective resolution $X \to k$ and a $k(G/H)$-projective resolution $Y \to k$. Also, choose diagonal chain maps

$$D_X : X \to X \otimes X$$
$$D_Y : Y \to Y \otimes Y$$

consistent with the diagonal homomorphisms of the groups. (We may even assume if necessary that these maps are co-associative and have co-units by choosing our resolutions properly.) Then $D_Y$ and $D_X$ induce a map

$$\text{Hom}_{G/H \times G/H}(Y \otimes Y, \text{Hom}_{H \times H}(X \otimes X, M \otimes N))$$
$$\to \text{Hom}_{G/H}(Y, \text{Hom}_H(X, M \otimes N)).$$

Moreover, we have an external product map

$$\text{Hom}_{G/H}(Y, \text{Hom}_H(X, M)) \otimes \text{Hom}_{G/H}(Y, \text{Hom}_H(X, N))$$
$$\to \text{Hom}_{G/H \times G/H}(Y \otimes Y, \text{Hom}_{H \times H}(X \otimes X, M \otimes N))$$

defined in the obvious way *but involving a sign*. (This map is even an isomorphism under often realizable assumptions on $X$ and $Y$, for example if both are free and finitely generated over $k$.) Putting these maps together, we obtain a pairing

$$\mathrm{Hom}_{G/H}(Y, \mathrm{Hom}_H(X, M)) \otimes \mathrm{Hom}_{G/H}(Y, \mathrm{Hom}_H(X, N))$$
$$\to \mathrm{Hom}_{G/H}(Y, \mathrm{Hom}_H(X, M \otimes N))$$

which provides a multiplicative structure on the relevant double complexes. Up to chain homotopy this multiplicative structure is associative, bigraded commutative, and has a unit. It is extremely tedious to do so, but one can verify that this multiplicative structure induces a product structure for each of the two spectral sequences discussed above. We have for each $r \geq 2$, a pairing

$$E_r(M) \otimes E_r(N) \to E_r(M \otimes N)$$

such that each $d_r$ is a derivation, and the multiplicative structure at each stage is just the cohomology of the multiplicative structure at the previous stage. Thus, a multiplicative structure is induced on $E_\infty$. Moreover, the product structure is consistent with the filtrations induced on the total cohomology of the double complexes in that $F^{p'} \otimes F^{p''} \to F^{p'+p''}$, and the isomorphism $E_\infty \cong Gr(H^*(A))$ preserves the multiplicative structure. In the case of the second spectral sequence, the induced map

$$H^{n'}(G, M) \otimes H^{n''}(G, N) \to H^{n'+n''}(G, M \otimes N)$$

is just the ordinary cup product. In the case of the first spectral sequence the product is defined as follows. The cup product $H^{q'}(H, M) \otimes H^{q''}(H, N) \to H^{q'+q''}(H, M \otimes N)$ is a pairing of $G/H$-modules, so it induces the ordinary cup product for $G/H$ cohomology

$$H^{p'}(G/H, H^{q'}(H, M)) \otimes H^{p''}(G/H, H^{q''}(H, N))$$
$$\to H^{p'+p''}(G/H, H^{q'+q''}(H, M \otimes N)).$$

The product for the $E_2$ term of the first spectral sequence is obtained by multiplying this ordinary cup product by the sign

$$(-1)^{q'p''}.$$

Suppose next that $L, M$, and $N$ are $kG$-modules, and we have a $kG$-homomorphism $M \otimes N \to L$. Then we may compose with the induced map to end up with products in the spectral structures associated with $L$. Suppose now that $H$ is a normal subgroup of $G$, and $M$ is a $G$-module on which $H$

acts *trivially*. Without loss of generality, we may take $k = \mathbf{Z}$. Consider the pairing
$$H/H' \otimes \mathrm{Hom}(H/H', M) \to M$$
obtained by evaluating homomorphisms on elements. It is easy to check that it is a $G$-homomorphism. Thus, it induces a pairing of spectral sequences as above. Consider first the pairing
$$E_2^{0,1}(H/H') \otimes E_2^{0,0}(\mathrm{Hom}) \to E_2^{0,1}(M)$$
or after suitable translation
$$\mathrm{Hom}_G(H/H', H/H') \otimes \mathrm{Hom}_G(H/H', M) \to \mathrm{Hom}_G(H/H', M).$$

Since this must be just the ordinary cup product, it is not hard to check that it is just composition of functions. In particular if $\iota \in \mathrm{Hom}_G(H/H', H/H')$ is the identity, we have
$$\iota\alpha = \alpha \qquad \text{for any } \alpha \in \mathrm{Hom}_G(H/H', M).$$

Thus $d_2(\alpha) = (d_2\iota)\alpha - \iota d_2\alpha = d_2\iota\alpha$. (Note that the two $\alpha$'s are in different places in the spectral sequence!) Thus, to compute $d_2^{0,1}$ in this case, it suffices to calculate $d_2\iota$. This can be done by a rather tedious explicit calculation using the bar resolution, and it turns out that
$$d_2\iota = -\epsilon$$
where $\epsilon \in H^2(G/H, H/H')$ is the class of the group extension
$$1 \to H/H' \to G/H' \to G/H \to 1.$$

Except for the explicit calculation (which we leave to the student), we have proved the following theorem.

**Theorem 7.3.1** (Hochschild–Serre). *Let $H$ be a normal subgroup of $G$ and let $M$ be a $G$-module on which $H$ acts trivially. Use the notation discussed above. Then for $\alpha \in H^1(H, M)^G = \mathrm{Hom}_G(H/H', M)$, we have $d_2\alpha = -\epsilon\alpha$.*

The identification of $d_2^{0,1}$ illustrates one of the drawbacks of the approach to group cohomology when developed purely in terms of homological algebra. The best way to understand the assertion $d_2\iota = -\epsilon$ is to prove it in the case of a *universal example*. Unfortunately, the universal example for this computation is a fibre space with base the Eilenberg–Mac Lane space $K(A, 2)$ and fibre $K(A, 1)$. The fibre may be dealt with by the algebraic theory, but the base and total space require use of a wider category such as the category of topological spaces.

Charlap and Vasquez (1966, 1969) have extended the above analysis to calculate $d_2$ in principle in many interesting cases. (See Evens and Friedlander (1982, Chapter III) for an application of their analysis.) It is useful to note that in some important cases, $d_2$ need not vanish even for a semi-direct product. (See the example below.)

**Exercise 7.3.1.** Let $G = K \ltimes H$ and $M$ a $G$ module on which $H$ acts trivially. Show that in the spectral sequence of the extension $H \to G \to K$, we have $d_2^{*,1} = 0$. (See also Proposition 7.3.2 below.)

The most important use we shall make of the multiplicative structure will be the case that $R$ is a ring on which $G$ acts as a group of ring automorphisms; then each $E_r(R)$ is a ring, $d_r$ is a derivation (with respect to *total* degree), etc. Similarly, if $R$ is such a ring, and if $M$ is a left $kG$-module which is also an $R$-module for which the structure morphism $R \otimes M \to M$ is a $G$-homomorphism, then each $E_r(M)$ is a left $E_r(R)$-module, each $d_r$ on $E_r(M)$ is a derivation in the extended sense, etc.

**Proposition 7.3.2.** *Let $G \cong H \rtimes K$ and suppose $M$ is a $G$-module on which $H$ acts trivially. Then* $\inf : H^*(G/H, M) \to H^*(G, M)$ *is a monomorphism onto a direct summand, and all differentials into the horizontal edge of the LHS spectral sequence for $M$ vanish.*

**Proof.** We may identify $K$ with $G/H$, and we have morphisms $\pi : G \to K$ and $\iota : K \to G$ such that $\pi\iota = \mathrm{Id}_K$. With the assumption $M^H = M$, these morphisms are consistent with $\mathrm{Id} : M \to M$. The asserted splitting follows functorially. In particular, it follows that the edge homomorphism $H^*(G/H, M) \cong E_2^{*,0} \to E_\infty^{*,0} \subseteq H^*(G, M)$ is a monomorphism so $E_2^{*,0} = E_r^{*,0} = E_\infty^{*,0}$ for $r \geq 2$. Since $E_{r+1}^{*,0} = E_r^{*,0}/\mathrm{Im}\, d_r^{*,r-1}$, it follows that $d_r^{*,r-1} = 0$. □

**Example.** Let $p$ be an odd prime, and let

$$G = Gp\langle a, b \mid a^{p^2} = 1, b^p = 1, bab^{-1} = a^{1+p}\rangle.$$

Let $H$ be the cyclic subgroup of order $p^2$ generated by $a$, and let $K$ be the cyclic subgroup of order $p$ generated by $b$. Then

$$G \cong H \rtimes K$$

is of order $p^3$.

Let $k$ be a field of characteristic $p$. Note that $K = G/H$ acts trivially on $H^q(H, k) \cong k$ because that is the only way that a $p$-group can act on $k$.

From the universal coefficient theorem (Section 3.4), it follows that
$$E_2 \cong H^*(K, H^*(H, k)) \cong H^*(K, k) \otimes H^*(H, k).$$
However, we know the structure of the cohomology ring of a cyclic group of odd order with coefficients in $k$:
$$H^*(K, k) = k[\beta_1, \beta_2 \mid \beta_1^2 = 0, \deg \beta_1 = 1, \deg \beta_2 = 2]$$
$$H^*(H, k) = k[\alpha_1, \alpha_2 \mid \alpha_1^2 = 0, \deg \alpha_1 = 1, \deg \alpha_2 = 2].$$
It follows that
$$E_2 = k[\beta_1, \beta_2, \alpha_1, \alpha_2]$$
where $\alpha_i \in E_2^{0,i}$, $\beta_i \in E_2^{i,0}$, $i = 1, 2$, satisfy the above relations, and they commute in pairs except for the relation
$$\alpha_1 \beta_1 + \beta_1 \alpha_1 = 0.$$
Because of the splitting, $d_2^{0,1} \alpha_1 = 0$. If $d_2^{0,2} \alpha_2 = 0$ also, it follows that $d_2 = 0$ in general since it vanishes on a set of generators for the ring $E_2$. However, in that case the same would be true for every $E_r$ by virtue of the position of the generators, so we would have $E_2 = E_\infty$. It would follow that
$$\dim H^2(G, k) = 1 + 1 + 1 = 3.$$
However, that conclusion is not correct (see below); hence we must have
$$d_2^{0,2} \alpha_2 = c \beta_2 \alpha_1$$
for some $c \neq 0 \in k$. From this we can calculate $d_2$ on any element, and thus determine $E_3$. Since $d_2$ is a derivation, $d_2(\alpha_2)^p = p(\alpha_2)^{p-1} d_2 \alpha_2 = 0$. Thus,
$$E_3 = k[\beta_1, \beta_2, \alpha_1, \alpha_1 \alpha_2, \ldots, \alpha_1 \alpha_2^{p-1}, \alpha_2^p].$$
Many products in this ring are zero either because of the previous relations or because they end up at zeros in the diagram. Since any differential into the horizontal edge is trivial, it follows by virtue of the position of the generators that $E_3 = E_\infty$. Thus, we have
$$E_\infty = k[\beta_1, \beta_2, \zeta_1, \zeta_3, \ldots, \zeta_{2p-1}, \zeta_{2p}]$$
where $\beta_i$ is at position $(i, 0)$ and $\zeta_j$ is at position $(0, j)$. Moreover,
$$\beta_1^2 = 0$$
$$\zeta_i \zeta_j = 0 \quad \text{for } i, j \text{ odd}$$
$$\beta_2 \zeta_j = 0 \quad \text{for } j = 1, \ldots 2p - 3$$
$$\beta_1 \zeta_j = -\zeta_j \beta_1 \quad \text{for } j = 1, \ldots, 2p - 1,$$

and all other generators commute.

To complete the argument, we must show that $\dim H^2(G,k) < 3$. The easiest way to do this is to calculate $H^2(G,\mathbf{Z})$ by means of the LHS spectral sequence. In this case, $H^1(H,\mathbf{Z}) = H^3(H,\mathbf{Z}) = 0$, and $K$ does not act trivially on $H^2(H,\mathbf{Z}) \cong \hat{H} \cong \mathbf{Z}/p^2\mathbf{Z}$. In fact, if $\alpha$ generates $\hat{H}$, then $b\alpha = (1-p)\alpha$. It is easy to check that $\hat{H}^K = p\mathbf{Z}\alpha$, and similarly for each of the other subgroups of $\hat{H} = \mathbf{Z}\alpha$ needed to calculate the cohomology of $K$ with coefficients in $\hat{H}$. It follows that $H^r(K, H^2(H,\mathbf{Z}))$ is cyclic of order $p$ for $r = 0$ and trivial for $r > 0$. Thus, $|H^2(G,\mathbf{Z})| = p^2$, and, since $H^2(K,\mathbf{Z})$ splits off, it follows that $H^2(G,\mathbf{Z}) \cong \mathbf{Z}/p\mathbf{Z} \oplus \mathbf{Z}/p\mathbf{Z}$. Since $H^3(K,\mathbf{Z}) = 0$, it follows that all factors in the filtration for $H^3(G,\mathbf{Z})$ are trivial, so $H^3(G,\mathbf{Z}) = 0$. Using the universal coefficient theorem or the short exact sequence

$$0 \to \mathbf{Z} \xrightarrow{p} \mathbf{Z} \to \mathbf{Z}/p\mathbf{Z} \to 0$$

gives the desired result for $H^2(G,\mathbf{F}_p)$ and hence for any field of characteristic $p$.

In general, $E_\infty$ only characterizes $H^*(G,k)$ modulo the filtration, and working out the latter from the former can be highly non-trivial. In this case, it is possible to show that $E_\infty \cong H^*(G,k)$ as rings. We present an outline of the argument with many details (which the reader should check) suppressed. Choose elements $\tilde{\zeta}_{2j-1} \in H^{2j-1}(G,k)$, $j = 1, 2, \ldots, p$, such that $\operatorname{res}_{G \to H} \tilde{\zeta}_{2j-1} = \alpha_1 \alpha_2^j$ and $\operatorname{res}_{G \to K} \tilde{\zeta}_{2j-1} = 0$. Then $\tilde{\zeta}_i$ represents $\zeta_i \in E_3^{0,i}$, and moreover it is uniquely determined by the specified conditions. (In those degrees, there are only two non-trivial quotients in the filtration and the lowest one splits off.) In addition, let $\tilde{\zeta}_{2p} = N_{H \to G}(\alpha_2)$ and put $\tilde{\zeta}_{2j-1} = \tilde{\zeta}_{2p} \tilde{\zeta}_{2j-2p-1}$ for $j > p$. Since $\operatorname{res}_{G \to H} \tilde{\zeta}_{2p} = \alpha_2^p$, it follows that $\tilde{\zeta}_i$ represents $\zeta_i$ for $i \geq 2p$. Finally, using the splitting, identify $\beta_1$ and $\beta_2$ with elements of $H^1(G,k)$ and $H^2(G,k)$ respectively.

Modulo the filtration, we have

$$\tilde{\zeta}_i \tilde{\zeta}_j = 0 \qquad \text{for } i, j \text{ odd}.$$

We claim that this is an exact equality in $H^*(G,k)$. To see this, consider the automorphism $\phi : G \to G$ defined by $\phi(a) = a^2, \phi(b) = b$. It is possible to check that at the level of cohomology

$$\phi^*(\beta_i) = \beta_i$$
$$\phi^*(\tilde{\zeta}_{2j-1}) = 2^j \tilde{\zeta}_{2j-1}$$
$$\phi^*(\tilde{\zeta}_{2p}) = 2\tilde{\zeta}_{2p}.$$

At worst, we can have (by virtue of positions in $E_\infty$)

$$\tilde\zeta_{2i-1}\tilde\zeta_{2j-1} = c\beta_1\tilde\zeta_{2(i+j-1)-1} \quad \text{for} \quad i+j \le p+1$$
$$= c\beta_1\tilde\zeta_{2(i+j-1)-1} + d\beta_2{}^{i+j-p-1}\tilde\zeta_{2p} + e\beta_1\beta_2{}^{i+j-p-1}\tilde\zeta_{2p-1}$$
$$\text{for} \quad p+1 < i+j \le 2(p-1).$$

With the exception of the case $i+j = 2(p-1)$ (i.e. $i = p$, $j = p-1$), it is easy to check that the only values of the constants consistent with the action of $\phi^*$ are 0. For the remaining case, the same argument shows that $c = 0$, multiplying by $\tilde\zeta_{2p-1}$ shows that $d = 0$, and multiplying by $\beta_2$ shows that $e = 0$. It follows that $\tilde\zeta_i\tilde\zeta_j = 0$ for $1 \le i, j \le 2p-1$. All the other relations are much easier to check using position arguments and noting the splitting off of the horizontal edge.

See Diethelm (1985) and Huebschmann (1989) for discussion of this and related examples.

**Exercise 7.3.2.**

(a) Fill in the details in the above determination of the ring structure of $H^*(G, k)$. In particular, verify the contentions about $\phi^*$. Hint: It suffices to determine the effect of $\phi$ on $H^*(H, k)$.

(b) Repeat the above analysis to determine $E_\infty$ for trivial coefficient module $\mathbf{Z}$. (See Wall (1961).)

(c) Show that the ring structure for $H^*(G, \mathbf{Z})$ is the same as that for $E_\infty$ in this case. Hint: Show that $\mathrm{cor}_{H \to G}(\alpha^i)$ generates $\mathrm{Ker}\{\mathrm{res}: H^{2i}(G, \mathbf{Z}) \to H^{2i}(K, \mathbf{Z})\}$ for $i = 1, \ldots, p-1$. Make use of the relation $\xi \, \mathrm{cor}_{H \to G}(\gamma) = \mathrm{cor}_{H \to G}(\mathrm{res}_{G \to H}(\xi)\gamma)$.

**Exercise 7.3.3.** Let $G = K \ltimes H$ with $|K|$ and $|H|$ relatively prime. Let $k$ be a commutative ring with $G$ acting trivially.

(a) Show that $H^*(G, k)$ is the direct product in the category of supplemented $k$-algebras of $H^*(K, k)$ and $H^*(H, k)^K$. Hint: The effect of taking the product in the category of supplemented $k$-algebras instead of the category of rings is to replace $k \times k$ by $k$ in degree 0. See Exercise 7.2.3.

(b) (Alperin–Atiyah) Let $K = \langle a \rangle \times \langle b \rangle$ where $a$ is of order 4 and $b$ is of order 2, and let $H = \langle x \rangle$ be cyclic of order 3. Define two homomorphisms of $K$ into $\mathrm{Aut}(H)$, one with kernel $\langle a \rangle$ and one with kernel $\langle a^2 \rangle \times \langle b \rangle$. Let $G_1$ and $G_2$ be the resulting semi-direct products. Show that $H^*(G_1, \mathbf{Z}) \cong H^*(G_2, \mathbf{Z})$ as rings but $G_1 \not\cong G_2$. (If the isomorphism between the integral cohomology rings is induced by a group homomorphism $\phi$, then $\phi$ must be a group isomorphism. See Evens (1961), where this result is stated without proof, and Jackowski (1978) or Benson and Evens (1990), where it is proved.)

## 7.4 Finiteness theorems

Recall that if $A$ is a (possibly non-commutative) ring, a left module $L$ over $A$ is called *noetherian* if the ascending chain condition applies to submodules of $L$. If so, then every submodule of $L$ as well as every quotient module is finitely generated. If $A$ is a graded ring, and $L$ is a graded module with no non-trivial elements of negative degree, then it is not hard to see that it does not make any difference whether we look at all submodules or only graded submodules. (See, for example, the proof of Lemma 7.4.5.) The rings we shall be interested in are either commutative or *commutative graded* (with graded modules), in which case it is not really necessary to distinguish between left and right modules. A commutative ring or commutative graded ring is noetherian if it is noetherian as a module over itself. If a commutative graded ring $A$ with no non-trivial elements of negative degree is noetherian, then the ideal of elements of positive degree is finitely generated, and it is easy to see that $A$ is finitely generated as a ring over $A^0$. A straightforward generalization of the *Hilbert Basis Theorem* asserts that if $A$ is a commutative ring, and $M$ is a noetherian module over $A$, then $A[X] \otimes_A M$ is a noetherian module over $A[X]$. (See for example Atiyah and MacDonald (1969, Chapter 7, Theorem 7.5 and Exercise 10).)

**Theorem 7.4.1.** *Let $G$ be a finite group, $k$ a commutative ring on which $G$ acts trivially, and $M$ a $kG$-module. If $M$ is noetherian as a $k$-module, then $H^*(G, M)$ is noetherian over $H^*(G, k)$.*

**Proof.** Let $H^{ev}(G, k)$ denote the subring of $H^*(G, k)$ of elements of *even* degree. $H^{ev}(G, k)$ is a commutative ring, and this will simplify some of the arguments. We shall in fact prove the slightly stronger assertion that $H^*(G, M)$ is noetherian over $H^{ev}(G, k)$.

*Reduction to the case of a p-group.*

Let $H^+(G, M)$ denote the submodule of $H^*(G, M)$ of elements of positive degree. We have $H^*(G, M)/H^+(G, M) \cong M^G$ where the action of $H^{ev}(G, k)$ on $M^G$ factors through $H^{ev}(G, k) \to k$. Since $M$ is noetherian over $k$, so is $M^G$. Hence, it suffices to show that $H^+(G, M)$ is noetherian. $H^+(G, M)$ breaks up as a finite direct sum of its $p$-primary components $H^+(G, M; p)$ for $p \mid |G|$. It is not hard to see that each $H^+(G, M; p)$ is an $H^{ev}(G, k)$-submodule of $H^+(G, M)$, so it suffices to show that each $H^+(G, M; p)$ is noetherian. Let $P$ be a $p$-Sylow subgroup of $G$. By corestriction theory (Proposition 4.2.5), if we put

$$r(k) = \text{Im}\{\text{res} : H^{ev}(G, k) \to H^{ev}(P, k)\}$$

and

$$t(k) = \bigoplus_{n \text{ even}, n > 0} \text{Ker}\{\text{cor} : H^n(P, k) \to H^n(G, k)\}$$

then
$$H^{ev}(P,k) = r(k) \oplus t(k).$$

(For elements of degree 0, this uses the fact that since $G$ acts trivially on $k$, res : $H^0(G,k) = k \to H^0(P,k) = k$ is the identity isomorphism.) Similarly,

$$H^+(P,M) = R^+(M) \oplus T(M)$$

where

$$R^+(M) = \bigoplus_{n>0} \text{Im}\{\text{res} : H^n(G,M) \to H^n(P,M)\}$$

$$T(M) = \bigoplus_{n>0} \text{Ker}\{\text{cor} : H^n(P,M) \to H^n(G,M)\}.$$

Clearly, it suffices to show that $R^+(M) \cong H^+(G,M;p)$ is noetherian over $r(k)$. We use the following lemma which follows easily from the formal properties of restriction and corestriction.

**Lemma 7.4.2.** $r(k)$ is a subring of $H^{ev}(P,k)$, and

$$r(k)R^+(M) \subseteq R^+(M), \quad t(k)R^+(M) \subseteq T(M).$$

**Proof.**

**Exercise 7.4.1.** Prove the lemma. Hint: Use the fact that restriction preserves cup products and $\text{cor}(\alpha \, \text{res} \, \beta) = \text{cor}(\alpha)\beta$. $\square$

Suppose that the theorem has been proved for $p$-groups so that we may assume $H^+(P,M)$ is noetherian over $H^{ev}(P,k)$. Let

$$J_1 \subseteq J_2 \subseteq \cdots \subseteq J_n \subseteq \cdots$$

be an ascending chain of $r(k)$-submodules of $R^+(M)$. Then

$$H^{ev}(P,k)J_n = (r(k) \oplus t(k))J_n = r(k)J_n \oplus t(k)J_n = J_n \oplus t(k)J_n$$

is the $n$th term of an ascending chain of $H^{ev}(P,k)$-submodules of $H^+(P,M)$ which we may assume stops. Since $J_n \subseteq R^+(M)$, $t(k)J_n \subseteq T(M)$, and $R^+(M) \cap T(M) = 0$, it follows that the original sequence $J_n$ stops as required.

$G$ *is a p-group*

We proceed by induction on the order of $G$. The theorem is clearly true if $|G| = 1$.

Let $Z$ be a non-trivial cyclic subgroup of $G$ contained in its centre. We consider the LHS spectral sequence

$$H^*(G/Z, H^*(Z, M)) \Rightarrow H^*(G, M)$$

as a module over the LHS spectral sequence

$$H^*(G/Z, H^*(Z, k)) \Rightarrow H^*(G, k).$$

Since $Z$ is central, $G/Z$ acts trivially on $H^*(Z, k)$. We know $H^{ev}(Z, k) = k[\chi]$ where $\deg \chi = 2$ and we may identify $\chi$ as an element of

$$E_2^{0,2}(k) = H^0(G/Z, H^2(Z, k)) = H^2(Z, k) = k\chi.$$

Since $G$ acts trivially on $\chi$, Theorem 6.1.1 (N4) tells us that $\chi^{(G:Z)} = \operatorname{res}_{G \to H}(N_{H \to G}(\chi))$ so $\zeta = \chi^{(G:Z)} \in E_\infty^{0,ev}(k)$. (In Proposition 7.2.2 we identified $E_\infty^{0,*}$ with the image of the restriction in $E_2^{0,*} = H^*(Z, M)^G$.) In particular, $d_r(\zeta) = 0$ for $r > 0$. Note that this statement makes sense since, on the vertical edge, $E_\infty^{0,*}$ is actually a subset of $E_2^{0,*}$.

Let $S = H^{ev}(G/Z, k) = E_2^{ev,0}$. We consider the subring $R = S[\zeta]$ of $E_2(k)$. Since $d_2$ vanishes on the horizontal edge, it follows that $d_2(R) = 0$ so $R$ projects onto a subring of $E_3(k) = H(E_2(k), d_2)$. By a similar argument, $R$ projects onto a subring of $E_r(k)$ for every $r$ including $r = \infty$. It follows that we may consider $E_r(M)$ as a module over $R$ for every $r$ including $r = \infty$.

**Lemma 7.4.3.** $E_2(M)$ *is a noetherian module over* $R$.

**Proof.** For $s > 0$, multiplication by $\chi \in H^2(Z, k)$ is an isomorphism

$$m_\chi : H^s(Z, M) \to H^{s+2}(Z, M).$$

Consider the cup product

$$H^0(G/Z, H^2(Z, k)) \otimes H^r(G/Z, H^s(Z, M)) \to H^r(G/Z, H^{s+2}(Z, M))$$

in the $E_2$ term of the spectral sequence. On the left,

$$\chi \in H^0(G/Z, H^2(Z, k)) = H^2(Z, k),$$

and it is not hard to see that multiplication by $\chi$ in the $E_2$ term is just the induced map

$$m_\chi^* : H^r(G/Z, H^s(Z, M)) \to H^r(G/Z, H^{s+2}(Z, M))$$

so that multiplication by $\chi$ is also an isomorphism on $E_2^{r,s}$ for $s > 0$. Hence the corresponding assertion also holds for multiplication by $\zeta$ except that the vertical degree is raised by $2(G:Z)$. Let

$$L = \bigoplus_{0 < s \leq 2(G:Z)} E_2^{*,s}(M).$$

Thus,

$$E_2^{*,+}(M) = \bigoplus_{j \geq 0} \zeta^j L$$

which may be rewritten

$$E_2^{*,+}(M) \cong S[X] \otimes_S L$$

where the action of $\zeta$ on the left corresponds to the action of $X$ on the right. However, each $H^s(Z, M)$ is a subquotient of $M$, so it is noetherian over $k$. Hence, by induction each $H^*(G/Z, H^s(Z, M))$ is noetherian over $S = H^{ev}(G/Z, k)$. Thus, $L$ is noetherian over $S$, so by the Hilbert Basis Theorem, $S[X] \otimes_S L$ is noetherian over $S[X]$. From the above remarks, it follows that $E_2^{*,+}(M)$ is noetherian over $R = S[\zeta]$ (which is an epimorphic image of $S[X]$). Since

$$E_2(M)/E_2^{*,+}(M) \cong E_2^{*,0}(M) \cong H^*(G/Z, M^Z)$$

with $S[\zeta]$ acting through the projection, $S[\zeta] \to S$ defined by sending $\zeta$ to $0$, it follows that $E_2(M)$ is noetherian over $R$ as claimed. $\square$

**Lemma 7.4.4.** *The spectral sequence for $M$ stops, i.e. $E_r(M) = E_\infty(M)$ for some $r < \infty$.*

**Proof.** Consider the 'pull back' $B_r$ in $E_2(M)$ of $d_r(E_r(M))$. This is defined as follows. Each element of $E_2(M)$ on which $d_2$ vanishes determines an element of $E_3(M)$. Suppose that $d_3$ vanishes on that element so that it in turn determines an element of $E_4(M)$. Continue placing restrictions in this way until you determine an element of $E_r(M)$, and suppose that element is in the image of $d_r$. The set of such elements in $E_2(M)$ is $B_r$. It is not hard to see that $B_r$ is an $R$-submodule since $d_j$ is a derivation for $2 \leq j \leq r$ and the image in each $E_j(k)$ of $R$ consists of universal cycles. Also, it is not hard to see that $B_r \subseteq B_{r+1}$ so that the $B_r$ form an ascending chain of $R$-submodules which must stabilize, say for $r \geq r_0$. It follows from this that $d_r = 0$ for $r > r_0$, and $E_r(M) = E_\infty$ for $r > r_0$. $\square$

We are now ready to complete the proof of Theorem 7.4.1.

Each $E_r(M)$ is a submodule of a quotient module of $E_{r-1}(M)$ so it follows by Lemma 7.4.3 and induction on $r$ that all the $E_r(M)$ are noetherian over $R$. Hence, by Lemma 7.4.4, $E_\infty(M)$ is noetherian over $R$. Since $R$ acts on $E_\infty(M) = Gr(H^*(G, M))$ through its homomorphic image in $Gr(H^{ev}(G, k)) = E_\infty^{ev}(k)$ (the subring of elements of even *total* degree), it follows that $Gr(H^*(G, M))$ is noetherian over $Gr(H^{ev}(G, k))$. The theorem now follows from the following lemma.

**Lemma 7.4.5.** *Let $A$ be a filtered graded ring, i.e. suppose*

$$A = F^0 A \supseteq F^1 A \supseteq \cdots \supseteq F^n A \supseteq \ldots$$

*where $F^p A \, F^q A \subseteq F^{p+q} A$, and*

$$F^p A = \bigoplus_n F^p A \cap A^n.$$

*Let $N$ be a filtered $A$-module, i.e. suppose*

$$N = F^0 N \supseteq F^1 N \supseteq \ldots F^n N \supseteq \ldots$$

*where $F^p A \, F^q N \subseteq F^{p+q} N$, and*

$$F^p N = \bigoplus_n F^p N \cap N^n.$$

*Suppose that $N^n = 0$ for $n < 0$, and for each $n$, $F^p N^n = F^p N \cap N^n = 0$ for $p$ sufficiently large (but possibly depending on $n$). If $Gr(N)$ is noetherian over $Gr(A)$, then $N$ is noetherian over $A$.*

**Proof.** This is basically a standard result in commutative algebra. (See for example, Bourbaki (1972, Chapter III, Section 2.9, Corollary 1).) We include the proof for completeness.

We want to show that the ascending chain condition (ACC) holds for submodules of $N$. First reduce to the case of *homogeneous* submodules as follows. If $L$ is a submodule of $N$, let $\tilde{L} = \oplus L^n$ where $L^n$ consists of 0 and all elements in $N^n$ which are *highest degree components* of elements in $L$. It is not hard to see that $\tilde{L}$ is a homogeneous submodule of $N$. Also, it is not hard to see that if $L' \subseteq L''$ and $\tilde{L}' = \tilde{L}''$, then $L' = L''$. That is enough to establish the reduction. For homogeneous submodules, the proof reduces similarly to showing that if $L' \subseteq L''$ and $Gr(L') = Gr(L'')$, then $L' = L''$. This follows easily by showing inductively that $L'/F^p L' = L''/F^p L''$ for each $p$, and then noting that, in any homogeneous component, $F^p L'^n = 0$ for $p$ sufficiently large. Also, one must use the fact that if $L$ is homogeneous, then $F^p L = F^p N \cap L = \oplus F^p N^n \cap L = \oplus F^p L^n$.  □

*Notes.*

Heretofore, we have envisioned a base ring over which everything is defined. Thus, we would form resolutions, use the Künneth Theorems over $k$, etc., all over this base ring. The ring $k$ in the statement of the theorem need not actually be this base ring; it could be an algebra over it. In fact, the theorem is even true if $k$ is not necessarily commutative but $G$ acts trivially, or if $G$ acts non-trivially and $k$ is a finitely generated algebra over a commutative noetherian base ring on which $G$ does act trivially. An interesting example of the latter would be a cohomology ring of a group on which $G$ acts. (See Evens (1961) for details.)

Lemma 7.4.4 showed that the LHS spectral sequence stops for a central extension with cyclic kernel. It is in fact true—subject to reasonable finiteness assumptions on the module—that the LHS spectral sequence for a finite group always stops. (See Evens (1975).)

**Corollary 7.4.6** (Venkov–Evens). *Let $G$ be a finite group and let $k$ be a noetherian ring on which $G$ acts trivially. Then $H^*(G,k)$ is a finitely generated $k$-algebra.*

**Proof.** $H^*(G,k)$ is a noetherian ring so $H^+(G,k)$ is a finitely generated ideal. By the usual argument, a set of ideal generators is also a set of generators for $H^*(G,k)$ as an algebra over $k$. □

The corollary was published by Venkov (1959) for $k = \mathbf{Z}$ or $\mathbf{F}_p$. His proof—which is quite simple—depends on imbedding the group $G$ in a unitary group $U(n)$, and then using basic facts about the cohomology of the classifying space $BU(n)$. The more general result stated in the theorem was discovered independently but appeared somewhat later in Evens (1961). It is possible by using sheaf theory to derive the generalization to modules by Venkov's geometric method, but we do not know of such a proof in the literature.

**Corollary 7.4.7.** *Let $G$ be a finite group, $H$ a subgroup, $k$ a ring on which $G$ acts trivially, and $M$ a $kH$-module which is noetherian over $k$. Then res : $H^{ev}(G,k) \to H^{ev}(H,k)$ induces an action of $H^{ev}(G,k)$ on $H^*(H,M)$ under which the latter is a noetherian module. In particular, if $k$ is a noetherian ring, then $H^*(H,k)$ is a finitely generated $H^{ev}(G,k)$-module.*

**Proof.** By the Eckmann–Shapiro Lemma (Proposition 4.1.3), there is an $H$-morphism
$$j : \mathrm{Hom}_H(kG, M) \to M$$
which induces an isomorphism
$$J^* = (j^* \, \mathrm{res}) : H^*(G, \mathrm{Hom}_H(kG, M)) \to H^*(H, M).$$
Given the stated action of $H^{ev}(G,k)$ on $H^*(H,M)$ and the fact that
$$j^* : H^*(H, \mathrm{Hom}_H(kG, M)) \to H^*(H, M)$$
is an $H^{ev}(H,k)$-module homomorphism, it follows that $J^*$ is an $H^{ev}(G,k)$-module isomorphism, and the corollary follows from the theorem since $\mathrm{Hom}_H(kG, M)$ is a finite sum of copies of $M$ as a $k$-module. □

# 8
# Varieties and complexity

## 8.1 The variety of a module

We shall use some basic facts about the prime ideal spectrum of a commutative ring. Good sources for this material are Bourbaki (1972) and Atiyah and MacDonald (1969). The latter text summarizes most of the relevant facts in exercises.

In what follows, unless otherwise asserted, $k$ is a field of characteristic $p$ and $G$ is a finite group. Define

$$H(G) = H(G,k) = \begin{cases} H^{ev}(G,k) & \text{for } p \text{ odd} \\ H^*(G,k) & \text{for } p = 2 \end{cases}$$

and let $X_G$ denote the topological space underlying the *prime ideal spectrum* $\operatorname{spec}(H(G,k))$. (Recall that the set of prime ideals of any commutative ring $R$ becomes a topological space under the Zariski topology. Technically, the notation $\operatorname{spec}(R)$ should be reserved for the *affine scheme* associated with $R$, so we shall distinguish the topological space—which is all that will concern us here—by the notation $|\operatorname{spec}(R)| = |\operatorname{spec}(R/\operatorname{rad} R)|$.) $X_G$ will be called the *variety* of the group $G$.

In this section and in what follows, the term rad for a ring or ideal will refer to the so-called *nil radical*. However, for a finitely generated commutative algebra over a field, the nil radical is the same as the Jacobson radical. (See Atiyah and MacDonald (1969, Chapter 5, Exercise 24).)

Suppose that $M$ and $N$ are finitely generated $kG$-modules. If $U \to k$ is a $kG$-projective resolution, then it follows that $U \otimes M \to k \otimes M = M$ is a $kG$-projective resolution of $M$. For it is certainly acyclic, so all we have to check is that $U \otimes M$ is projective if $U$ is. This follows easily from the case $U = kG$ in which we can exhibit an explicit isomorphism from $kG \otimes M$ to the $kG$-free module $kG \otimes \tilde{M}$ where $\tilde{M}$ is the vector space $M$ with trivial $G$-action. Namely, let $x \otimes m \mapsto x \otimes x^{-1}m$. This defines an isomorphism (with inverse defined by $x \otimes m \mapsto x \otimes xm$), and it is easy to check that it is a $G$-homomorphism $kG \otimes M \to kG \otimes \tilde{M}$. (Compare with the proof of Lemma 7.2.1.)

Since $\operatorname{Hom}_G(U \otimes N, M) \cong \operatorname{Hom}_G(U, \operatorname{Hom}_k(N, M))$ it follows that there is a natural equivalence of functors

$$\operatorname{Ext}^*_{kG}(N, M) \cong H^*(G, \operatorname{Hom}(N, M)).$$

(If $k = \mathbf{Z}$, then a simple spectral sequence argument (Exercise 7.2.1) shows that there is a long exact sequence

$$\cdots \to H^n(G, \operatorname{Hom}_\mathbf{Z}(N, M)) \to \operatorname{Ext}^n_{\mathbf{Z}G}(N, M)$$
$$\to H^{n-1}(G, \operatorname{Ext}^1_\mathbf{Z}(N, M)) \to H^{n+1}(G, \operatorname{Hom}_\mathbf{Z}(N, M)) \to \cdots$$

so the situation is a bit more complicated.)

Since, under the above circumstances, $\operatorname{Hom}(N, M)$ is a finite dimensional $k$-space, it follows that we may view $\operatorname{Ext}^*_{kG}(N, M)$ as a finitely generated module over the noetherian ring $H(G, k)$. We shall usually obtain this module structure through the above isomorphism, but it could also be defined directly as a composition

$$\operatorname{Ext}^*_G(k, k) \otimes \operatorname{Ext}^*_G(N, M) \to \operatorname{Ext}^*_{G \times G}(k \otimes N, k \otimes M) \to \operatorname{Ext}^*_G(k \otimes N, k \otimes M)$$

in essentially the same way as we defined the cup product for $H^*(G)$ in the first place. Consider, in particular, the case $M = N$. Then $\operatorname{Hom}(M, M)$ is a finite dimensional $k$-algebra, and $k \to \operatorname{Hom}(M, M)$ defined by $1 \mapsto \operatorname{Id}$ induces a ring homomorphism

$$H^*(G, k) \to H^*(G, \operatorname{Hom}(M, M)) = \operatorname{Ext}^*_{kG}(M, M)$$

and $H(G, k)$ is carried into the centre of $\operatorname{Ext}^*_{kG}(M, M)$. To see this, consider the consequences of the diagram

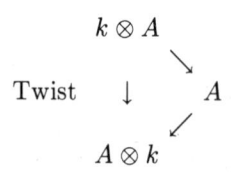

$$\begin{array}{ccc} & k \otimes A & \\ & & \searrow \\ \text{Twist} & \downarrow & A \\ & & \nearrow \\ & A \otimes k & \end{array}$$

where $A = \operatorname{Hom}(M, M)$.

Let

$$\mathfrak{a}_G(M) = \operatorname{Ker}\{H(G, k) \to H^*(G, \operatorname{Hom}(M, M))\}$$
$$= \operatorname{ann}_{H(G)}(\operatorname{Ext}^*_{kG}(M, M))$$

and let $I_G(M)$ be the radical of $\mathfrak{a}_G(M)$. Set

$$X_G(M) = V(\mathfrak{a}_G(M)) = |\operatorname{spec}(H(G)/\mathfrak{a}_G(M))| = |\operatorname{spec}(H(G)/I_G(M))|.$$

(If $\mathfrak{a}$ is an ideal, $V(\mathfrak{a})$ is the closed subspace of $X_G$ defined by $\mathfrak{a}$, i.e. the set of prime ideals $\mathfrak{p} \supseteq \mathfrak{a}$.) $X_G(M)$ is called the *variety of the module* $M$. It may also be identified as the set of points $\mathfrak{p} \in X_G$ such that the localization $H^*(G, \text{Hom}(M,M))_\mathfrak{p} \neq 0$, i.e. it is the *support* of the $H(G)$-module $H^*(G, \text{Hom}(M,M)) = \text{Ext}^*_G(M,M)$.

Clearly, $X_G(k) = X_G$.

More generally, we could define analogous objects for $\text{Ext}^*_{kG}(N,M)$ for any pair of $kG$-modules $M, N$. We shall show that for fixed $M$ we obtain *the same result* for any $N$. To this end, consider the $kG$-pairing

$$\text{Hom}(M,M) \otimes \text{Hom}(N,M) \to \text{Hom}(N,M)$$

given by $g \otimes f \mapsto g \circ f$. Through the cup product, this induces an $H(G,k)$-pairing

$$H^*(G, \text{Hom}(M,M)) \otimes H^*(G, \text{Hom}(N,M)) \to H^*(G, \text{Hom}(N,M))$$

which we denote $\alpha \circ \beta$. Then any $\alpha \in \text{Ext}^*_{kG}(N,M)$ can be written $\alpha = \text{Id} \circ \alpha$ where $\text{Id} \in \text{Ext}^0_{kG}(M,M) = \text{Hom}_G(M,M))$. Thus, for $\xi \in \mathfrak{a}_G(M)$, we have

$$\xi \text{Ext}^*_G(N,M) = \xi(\text{Id} \circ \text{Ext}^*_{kG}(N,M))$$
$$= (\xi \text{ Id}) \circ H^*(G, \text{Hom}(N,M)) = 0$$

for any $N$. Conversely, if $\xi$ annihilates $\text{Ext}^*_{kG}(N,M)$ for every $N$, then it certainly annihilates $\text{Ext}^*_{kG}(M,M)$.

Clearly, the same argument applies with the roles of $M$ and $N$ reversed. Thus we have proved the following result.

**Proposition 8.1.1.** *Let $M$ be a finitely generated $kG$-module, and let $\xi \in H(G)$. Then the following are equivalent.*
 (i) $\xi \in \mathfrak{a}_G(M)$.
 (ii) $\xi \text{Ext}^*_{kG}(N,M) = 0$ *for every* $N$.
 (iii) $\xi \text{Ext}^*_{kG}(M,N) = 0$ *for every* $N$.
 (iv) $\xi \text{ Id} = 0$ *for* $\text{Id} \in \text{Hom}_G(M,M) \subseteq \text{Ext}^*_{kG}(M,M)$.

The corresponding result for the radical of $\mathfrak{a}_G(M)$ is sometimes useful.

**Corollary 8.1.2.** *Let $M$ be a finitely generated $kG$-module. Then $I_G(M)$ is the set of all $\xi \in H(G)$ such that there is a $j > 0$ with $\xi^j \text{Ext}^*_G(N,M) = 0$ (alternatively $\xi^j \text{Ext}^*_G(M,N) = 0$) for every finitely generated $kG$-module $N$.*

*Notes.*

1. The proposition says that $X_G(M)$ is the union of the supports of the $H(G)$-modules $\text{Ext}^*_G(M,N)$ — in essence its original definition. (See Alperin and Evens (1981).) The observation that one could use the module $\text{Hom}(M,M)$ is due to Carlson (1983).

2. The product structure on $\operatorname{Ext}_{kG}(-,-)$ induced by the pairing

$$\operatorname{Hom}(M, N) \otimes \operatorname{Hom}(L, M) \to \operatorname{Hom}(L, N)$$

for $k$ a field is same as the *Yoneda product* $\alpha \circ \beta$. (See Mac Lane (1963, Section III.5) or (Hilton and Stammbach 1971, Section IV.9) for a general discussion of Yoneda theory.) For more general $k$ there are cup product pairings

$$H^*(G, k) \otimes \operatorname{Ext}_{kG}^*(M, N) \to \operatorname{Ext}_{kG}^*(M, N)$$
$$\operatorname{Ext}_{kG}^*(M, N) \otimes H^*(G, k) \to \operatorname{Ext}_{kG}^*(M, N).$$

It is possible to show that, for $\alpha \in \operatorname{Ext}_{kG}^p(M, N)$, $\beta \in \operatorname{Ext}_{kG}^q(L, M)$, and $\chi \in H^r(G, k)$,

$$\chi(\alpha \circ \beta) = (\chi\alpha) \circ \beta$$
$$(\alpha \circ \beta)\chi = \alpha \circ (\beta\chi).$$

3. In the current notation, the original definition of the variety of $M$ gives $X_G(M^*)$ where $M^* = \operatorname{Hom}(M, k)$. Fortunately, we have the following proposition.

**Proposition 8.1.3.** *For every finitely generated $kG$-module $M$, we have*

$$X_G(M) = X_G(M^*).$$

**Proof.** Sending $f : M \to M$ into its dual $f^* : M^* \to M^*$ provides a linear isomorphism

$$\operatorname{Hom}(M, M) \cong \operatorname{Hom}(M^*, M^*)$$

(in fact an anti-isomorphism of algebras) which can be seen to be a $kG$-map. Hence, the two cohomology modules have the same support. □

**Proposition 8.1.4.** *Suppose $0 \to M' \to M \to M'' \to 0$ is a short exact sequence of finitely generated $kG$-modules. Then*

$$X_G(M) \subseteq X_G(M') \cup X_G(M''),$$

*and a similar relation applies for any permutation of $\{M', M, M''\}$. If the sequence splits, then*

$$X_G(M) = X_G(M') \cup X_G(M'').$$

**Proof.** For each finitely generated $kG$-module $N$,

$$\operatorname{Ext}_{kG}^*(N, M') \to \operatorname{Ext}_{kG}^*(N, M) \to \operatorname{Ext}_{kG}^*(N, M'')$$

is exact. Choose $j'$ such that $\xi^{j'} \operatorname{Ext}_{kG}^*(N, M') = 0$ for every finitely generated $kG$-module $N$. Similarly, choose $j''$ such that $\xi^{j''} \operatorname{Ext}_{kG}^*(N, M'') = 0$ for every such $N$. Then $\xi^{j''} \operatorname{Ext}_{kG}^*(N, M)$ is in the image of $\operatorname{Ext}_{kG}^*(N, M')$, and
$$\xi^{j'+j''} \operatorname{Ext}_{kG}^*(N, M) = 0$$
for every such $N$. Hence,
$$I_G(M) \supseteq I_G(M') \cap I_G(M'')$$
and the result follows from the fact that $V(-)$ inverts inclusions and carries intersections to unions.

According to Proposition 3.3.2 the connecting homomorphism
$$\operatorname{Ext}_{kG}(N, M'') \to \operatorname{Ext}_{kG}(N, M')$$
is a morphism of $H(G)$-modules, so we may mimic the above argument for any permutation of the modules $M', M, M''$.

The case of the direct sum $M = M' \oplus M''$ is left as an exercise for the reader. □

The argument in the proof of Proposition 8.1.4 actually shows the following.

**Corollary 8.1.5.** $\xi \in I_G(M)$ if and only if for each simple module $N$,
$$\xi^j \operatorname{Ext}_{kG}^*(N, M) = 0$$
for some $j > 0$. A similar result holds using instead
$$\xi^j \operatorname{Ext}_{kG}^*(M, N) = 0.$$

**Proposition 8.1.6.** If $G$ is a $p$-group and $M$ is a finitely generated $kG$-module, then $X_G(M)$ is the support of the $H(G)$-module $H^*(G, M)$.

**Proof.** If $G$ is a $p$-group, the only simple $kG$-module is $k$ so we are reduced to considering $\operatorname{Ext}_G^*(k, M) = H^*(G, M)$. □

## 8.2  Subgroups

If $H$ is a subgroup of $G$, then we denote by $\iota_{H \to G} : X_H \to X_G$ the map induced by the ring homomorphism $\operatorname{res}_{G \to H} : H(G) \to H(H)$. If the context makes the meaning clear, we shall omit part or all of the subscript in $\iota_{H \to G}$.

By Corollary 7.4.7, $H(H)$ is a finite module over $\operatorname{Im}\operatorname{res} \cong H(G)/\operatorname{Ker}\operatorname{res}$, so by the Going-up Theorems for noetherian rings, the induced map

$$X_H \to |\operatorname{spec}\operatorname{Im}\operatorname{res}|$$

is surjective with *each point having a finite pre-image*. (Each prime ideal of $\operatorname{Im}\operatorname{res}$ is the intersection with a prime ideal in the over-ring $H(H)$ and the number of such prime ideals above it is finite. See Atiyah and Macdonald (1969, Chapter 8, Exercise 4).) It follows that

$$\iota_H(X_H) = V(\operatorname{Ker}\operatorname{res}) = |\operatorname{spec}(H(G)/\operatorname{Ker}\operatorname{res})| = |\operatorname{spec}\operatorname{Im}\operatorname{res}|$$

and by a similar argument for any $H$-module $M$,

$$\iota_H(X_H(M)) = V(\operatorname{res}^{-1}(I_H(M))) = |\operatorname{spec}(H(G)/\operatorname{res}^{-1}(I_H(M)))|.$$

Note that there are some *notational difficulties* in the above discussion. If $M$ is a $G$-module, it is usual in the literature on representation theory to denote the $H$-module obtained by restriction by $M_H$ or $M{\downarrow}_H^G$. Unfortunately, the first notation has been preempted in group homology, and the second notation is a bit cumbersome. Hence, as previously in this book, we shall just use the symbol $M$ and hope that the context will make clear what it is supposed to be a module over.

**Proposition 8.2.1.** *Let $M$ be a finitely generated $kG$-module. Then*

$$I_G(M) \subseteq \operatorname{res}^{-1}(I_H(M))$$

*i.e.*

$$X_G(M) \supseteq \iota_H(X_H(M)).$$

**Proof.** It suffices to show that if $\xi \in I_G(M)$, then some power of $\operatorname{res}\xi \in H(H)$ annihilates $\operatorname{Ext}^*_H(M, T)$ for every $kH$-module $T$. By the Eckmann–Shapiro Lemma, we have an $H(G)$-module isomorphism

$$H^*(G, \operatorname{Hom}(M, T){\uparrow}_H^G) \cong H^*(H, \operatorname{Hom}(M, T)).$$

However, for $M$ a $kG$-module, we have

$$\operatorname{Hom}(M, T){\uparrow}_H^G = \operatorname{Hom}_H(kG, \operatorname{Hom}(M, T))$$
$$\cong \operatorname{Hom}(M, \operatorname{Hom}_H(kG, T)) \cong \operatorname{Hom}(M, T{\uparrow}_H^G)$$

and some power of $\xi$ annihilates $H^*(G, \operatorname{Hom}(M, T{\uparrow}_H^G))$.

(Note that we could equivalently have directly invoked

$$\operatorname{Ext}^*_G(M, \operatorname{Hom}_H(kG, T)) \cong \operatorname{Ext}^*_H(M, T)$$

which is a form of the Eckmann–Shapiro Lemma for Ext mentioned in Section 4.1.) □

**Proposition 8.2.2.** *If $P$ is a p-Sylow subgroup of $G$ and $M$ is a finitely generated $kG$-module, then*
$$I_G(M) = \text{res}^{-1}(I_P(M)), \text{ i.e. } \iota_P(X_P(M)) = X_G(M).$$

**Proof.** We need only prove that $I_G(M) \supseteq \text{res}^{-1}(I_P(M))$. Let $\text{res}\,\xi \in I_P(M)$, i.e. assume some power of $\text{res}\,\xi$ annihilates $H^*(P, \text{Hom}(M,M))$. Since
$$H^*(G, \text{Hom}(M,M)) \to H^*(P, \text{Hom}(M,M))$$
is a monomorphism, it follows that some power of $\xi$ annihilates
$$H^*(G, \text{Hom}(M,M))$$
and $\xi \in I_G(M)$. □

**Proposition 8.2.3.** *Let $G$ be a finite group, $M$ a finitely generated $kG$-module, and $H$ a subgroup. For any $x \in G$, we have*
$$\iota_H(X_H(M)) = \iota_{xHx^{-1}}(X_{xHx^{-1}}(M))$$

**Proof.**

**Exercise 8.2.1.** Prove the proposition. □

**Proposition 8.2.4.** *Let $G$ be a finite group, $H$ a subgroup, and $M$ a finitely generated $kH$-module. Let $M\uparrow_H^G$ denote the induced $kG$-module. Then*
$$X_G(M\uparrow_H^G) = \iota_H(X_H(M)).$$

**Proof.** As noted at the beginning of this section,
$$\iota_H(X_H(M)) = V(\text{res}^{-1}(I_H(M))),$$
so it suffices to show that
$$I_G(M\uparrow_M^G) = \text{res}^{-1}(I_H(M)).$$
To see this, note, as in the proof of Proposition 8.2.1, that the isomorphism
$$\text{Ext}_G^*(M\uparrow_H^G, M\uparrow_H^G) \cong \text{Ext}_H^*(M\uparrow_H^G, M)$$
implies that $\rho \in I_G(M\uparrow_H^G)$ if and only if some power of $\text{res}\,\rho$ kills the right hand side. However, viewed as an $H$-module, $M\uparrow_H^G$ contains $M$ as a direct summand, so it follows that any such $\text{res}\,\rho \in I_H(M)$. By Corollary 8.1.2, it follows from $\text{res}\,\rho \in I_H(M)$ that some power kills $\text{Ext}_H^*(M\uparrow_H^G, M)$ so we obtain
$$\rho \in I_G(M\uparrow_H^G) \iff \text{res}\,\rho \in I_G(M)$$
as required. □

## 8.3  Relations with elementary abelian $p$-subgroups

For any elementary abelian $p$-subgroup $E$ of $G$,

$$H(E)/\operatorname{rad}(H(E)) \cong k[\xi_1, \ldots, \xi_d]$$

is a polynomial ring. Indeed, for $p = 2$, $H(E) = H^*(E, k)$ is already a polynomial ring, while for $p > 2$, since $H^*(E, k)$ is a polynomial ring tensored with an exterior algebra, every element of even degree which is not in the polynomial ring has square 0. Thus, we may view $X_E = |\operatorname{spec}(H(E))|$ as affine $d$-space over $k$. Moreover, $\operatorname{Im} \operatorname{res}_{G \to E}$ modulo its radical is a subring of a domain, so $\iota_E(X_E) = |\operatorname{spec}(H(G)/\operatorname{Ker} \operatorname{res})|$ is much easier to deal with.

**Theorem 8.3.1** (Alperin–Avrunin–Evens). *Let $G$ be a finite group, $k$ a field of characteristic $p$ and $M$ a finitely generated $kG$-module. Then*

$$X_G(M) = \bigcup_{\substack{E \leq G \\ E \text{ elem ab } p\text{-gp}}} \iota_E(X_E(M)).$$

(See Alperin and Evens (1982) and Avrunin (1981). The statement of the Theorem was suggested to Alperin and Evens by Serre.)

**Proof.** Fix a $p$-Sylow subgroup $P$ of $G$. Since every elementary abelian $p$-subgroup is conjugate to a subgroup of $P$, we may restrict attention to such $E$ by Proposition 8.2.3. Moreover, by Proposition 8.2.2, it suffices to prove the theorem for $P$, so we assume $G$ is a $p$-group.

Since $\iota_E(X_E(M)) = V(\operatorname{res}^{-1}(I_E(M)))$ and since $V(\mathfrak{a} \cap \mathfrak{b}) = V(\mathfrak{a}) \cup V(\mathfrak{b})$, we may rewrite the desired conclusion

$$I_G(M) = \bigcap_{\substack{E \leq G \\ E \text{ elem ab } p\text{-gp}}} \operatorname{res}^{-1}(I_E(M)).$$

$X_G(M)$ contains the union by Proposition 8.2.1, so it suffices to prove the opposite inclusion. In terms of ideals, we want to show that $I_G(M)$ contains the intersection of ideals on the right, i.e. we want to show the following: if for each $E$ some power of $\operatorname{res}_E(\xi)$ annihilates $\operatorname{Ext}^*_E(M, M)$, then some power of $\xi$ annihilates $\operatorname{Ext}^*_G(M, M)$. If the $p$-group $G$ is elementary abelian, the conclusion is obvious. Otherwise, every elementary abelian subgroup is contained in a maximal subgroup $H$ so we may assume for each such $H$ that some power of $\operatorname{res}_H(\xi)$ annihilates $\operatorname{Ext}^*_H(M, M)$.

For each maximal subgroup $H$, we have the LHS spectral sequence

$$H^*(G/H, \operatorname{Ext}^*_H(M, M)) \Rightarrow \operatorname{Ext}^*_G(M, M)$$

as a module over the spectral sequence

$$H^*(G/H, H^*(H, \mathbf{Z})) \Rightarrow H^*(G, \mathbf{Z}).$$

Such a spectral sequence may be derived along with all the requisite multiplicative structures by mimicking the construction we used for the original Lyndon–Hochschild–Serre spectral sequence in Section 7.2. However, in our case, this is not necessary since we may use the fact that

$$\operatorname{Ext}_G^*(M, M) = H^*(G, \operatorname{Hom}(M, M)).$$

Since $H$ is maximal, $G/H$ is cyclic of order $p$, and the $E_2$-term has a particularly simple multiplicative structure. Namely, write $G/H = P$ and choose $\beta$ generating $H^2(P, \mathbf{Z}) = \operatorname{Hom}(P, \mathbf{Q}/\mathbf{Z})$. Put $N = \operatorname{Hom}(M, M)$. Since $\beta$ is a universal cycle (i.e. $d_r\beta = 0$ for all $r$), multiplication by $\beta$ induces a homomorphism $\phi_r : E_r^{*,*}(N) \to E_r^{*+2,*}(N)$ for each $r \geq 2$. Let $\beta_H = \inf \beta \in H^2(G, \mathbf{Z})$, so $H = \operatorname{Ker} \beta_H$.

**Lemma 8.3.2** (Quillen–Venkov).

$$\phi_r^{s,*} : E_r^{s,*}(N) \to E_r^{s+2,*}(N)$$

*is an epimorphism for all $s \geq 0$ and is an isomorphism for $s \geq r$. As a consequence $F^{s+2} \operatorname{Ext}_G^*(M, M) = \beta_H F^s \operatorname{Ext}_G^*(M, M)$.*

**Proof.** The assertions about $\phi_r$ follow by checking on elements. They are true in $E_2$ by the known behaviour of multiplication for cohomology of cyclic groups. (The fact that we only obtain an epimorphism for $s = 0$ is due to the fact that $H^0(P, A) = A^P$ while $H^2(P, A) = A^P/TA$.) Suppose $\xi^{s+2,q} \in E_r^{s+2,q}$. Choose $\tilde{\xi}^{s+2,q} \in E_{r-1}^{s+2,q}$ representing it. By induction, we may assume $\tilde{\xi}^{s+2,q} = \beta \tilde{\xi}^{s,q}$. To show $\phi_r$ is an epimorphism, it suffices to show that $d_{r-1}\tilde{\xi}^{s,q} = 0$. However,

$$\phi_{r-1}(d_{r-1}\tilde{\xi}^{s,q}) = \beta d_{r-1}\tilde{\xi}^{s,q} = d_{r-1}\beta\tilde{\xi}^{s,q} = d_{r-1}\tilde{\xi}^{s+2,q} = 0.$$

Since $d_{r-1}\tilde{\xi}^{s,q} \in E_{r-1}^{s+r-1,q-r+2}$, we are in the range in which $\phi_{r-1}$ is an isomorphism, so the conclusion follows. The argument showing $\phi_r$ is a monomorphism for $s \geq r$ is similar, and is left as an exercise for the reader.

The last assertion follows by examining the $E_\infty$-term. For a fixed total degree $n$, we may proceed by decreasing induction on $s$. Since

$$F^{n+2}H^{n+2} = E_\infty^{n+2,0} = \beta E_\infty^{n,0} = \beta_H F^n H^n,$$

we can start the induction. However, a similar calculation for each

$$F^{s+2}H^{n+2}/F^{s+3}H^{n+2} = E_\infty^{s+2,n-s}$$

allows us to go from $s$ to $s+1$ where we can assume the result.
This completes the proof of the lemma. □

To continue with the proof of the theorem, let some power of $\zeta = \operatorname{res}_H(\xi) \in H(H,k)$ annihilate $\operatorname{Ext}^*_H(M,M)$. We may identify $\zeta$ with an element of
$$H^0(G/H, H^*(H,k)) = E_2^{0,*}(k).$$
However, if $A$ is any $G/H$-ring (e.g. $H^*(H,k)$) and $B$ is a compatible $G/H$-module (e.g. $\operatorname{Ext}^*_H(M,M)$), then the pairing
$$H^0(G/H, A) \otimes H^s(G/H, B) \to H^s(G/H, B)$$
is obtained simply by multiplication by invariant elements at the cochain level. Hence, for each $s \geq 0$, $H^s(G/H, \operatorname{Ext}^*_H(M,M)) = E_2^{s,*}$ is annihilated by some power of $\zeta$. Since $\zeta$ is a universal cycle ($\zeta = \operatorname{res} \xi$ for $\xi \in H(G,k)$), some power of $\zeta$ annihilates $E_\infty^{s,*}$. Taking $s = 0$, we have
$$\xi^i(\operatorname{Ext}^*_G(M,M)/F^1 \operatorname{Ext}^*_G(M,M)) = 0$$
for some $i > 0$, and taking $s = 1$, we have
$$\xi^j(F^1 \operatorname{Ext}^*_G(M,M)/F^2 \operatorname{Ext}^*_G(M,M)) = 0$$
for some $j > 0$. It follows that
$$\xi^{i+j} \operatorname{Ext}^*_G(M,M) \subseteq F^2 \operatorname{Ext}^*_G(M,M) = \beta_H \operatorname{Ext}^*_G(M,M)$$
by the lemma. By Serre's Theorem (Theorem 6.4.1), choose maximal subgroups $H_1, \ldots, H_n$ such that
$$\beta_1 \beta_2 \ldots \beta_n = 0$$
where $\beta_i = \beta_{H_i}$. Starting with $H = H_1$ above, we can find successive powers such that
$$\xi^{m_1} \operatorname{Ext}^*_G(M,M) \subseteq \beta_1 \operatorname{Ext}^*_G(M,M)$$
$$\xi^{m_2+m_1} \operatorname{Ext}^*_G(M,M) \subseteq \xi^{m_2}\beta_1 \operatorname{Ext} = \beta_1 \xi^{m_2} \operatorname{Ext} \subseteq \beta_1 \beta_2 \operatorname{Ext}$$
$$\vdots$$
$$\xi^{m_n+\cdots+m_1} \operatorname{Ext}^*_G(M,M) \subseteq \cdots \subseteq \beta_1 \ldots \beta_n \operatorname{Ext}^*_G(M,M) = 0$$
as required.

*Remark.* The Quillen–Venkov Lemma was originally stated in Quillen and Venkov (1972) for $M = k$, and in terms of the action of $H^*(P, \mathbf{F}_p)$

on the spectral sequence. In that case, we may take $\beta_H = \delta\alpha_H$ for some $\alpha_H \in H^1(G, \mathbf{F}_p) = \mathrm{Hom}(G, \mathbf{F}_p)$ with kernel $H$. For $p = 2$, the conclusion in fact holds with $\alpha_H$ replacing $\beta_H$ and we obtain $F^{s+1}\mathrm{Ext}^*(M,M) = \alpha_H F^s \mathrm{Ext}^*(M,M)$. We leave it to the reader to fill in the details. □

**Exercise 8.3.1.** Show $\phi_s$ is a monomorphism for $s \geq r$ in the Quillen–Venkov Lemma.

**Exercise 8.3.2.** For $p = 2$, show that

$$\mathrm{Ker}\,\mathrm{res}_{G \to H} = F^1 H^*(G, k) = \alpha_H H^*(G, k).$$

The case $M = k$ is the following theorem of Quillen (1971a, 1971b).

**Corollary 8.3.3** (Quillen). *Let $G$ be a finite group, and $k$ a field of characteristic $p$. Then*

$$X_G = \bigcup_{\substack{E \leq G \\ E \text{ elem ab } p\text{-gp}}} \iota_E(X_E)$$

*Also,*

$$\dim X_G = \max_{\substack{E \leq G \\ E \text{ elem ab } p\text{-gp}}} \dim X_E = \max_{\substack{E \leq G \\ E \text{ elem ab } p\text{-gp}}} \mathrm{rank}\,(E).$$

**Proof.** Recall that $\dim X_G$ is the dimension of $H(G)$ or $H(G)/\mathrm{rad}(H(G))$. The first assertion follows immediately from the theorem. The second assertion follows if we can show that $\dim \iota_E(X_E) = \dim X_E$. However, as noted earlier, $H(E)$ is a finite module over $\mathrm{Im}\,\mathrm{res}_{G \to E}$ so they have the same Krull dimension. □

**Corollary 8.3.4.** *Let $G$ be a finite group, and $k$ a field of characteristic $p$. $\xi \in H^*(G, k)$ is nilpotent if and only if $\mathrm{res}_E \xi$ is nilpotent for every elementary abelian p-subgroup $E$ of $G$.*

**Proof.** For $p = 2$ or for elements of even degree, this is just a restatement of the formula for $I_G(M)$ in the case $M = k$. If $p$ is odd, elements of odd degree are always nilpotent anyway. □

## 8.4 Complexity

Let $k$ be a field of characteristic $p$, $G$ a finite group, and $M$ a finitely generated $kG$-module. We define the *complexity* of the module $M$ by

$$cx_G(M) = \dim X_G(M) = \dim H(G)/\mathfrak{a}_G(M) = \dim H(G)/I_G(M).$$

**Theorem 8.4.1.** *The complexity of $M$ is zero if and only if $M$ is $kG$-projective.*

**Proof.** If $M$ is $kG$-projective, then $\operatorname{Ext}^n_{kG}(M,M) = 0$ for $n > 0$ so $\mathfrak{a}_G(M) = H^+$, the maximal ideal of elements of positive degree. Hence,

$$|\operatorname{spec} H(G)/\mathfrak{a}_G(M)|$$

is a point, and its dimension is 0.

Conversely, suppose $cx_G(M) = 0$. Then, $H(G)/\mathfrak{a}_G(M)$ is a noetherian ring of dimension 0, so it is artinian, and hence finite dimensional over $k$. (See Atiyah and MacDonald (1969, Chapter 8).) If $N$ is a finitely generated $kG$-module, $\operatorname{Ext}^*_{kG}(M,N)$ is a finite module over $H(G,k)$. However, by Proposition 8.1.1, $\mathfrak{a}_G(M)$ kills $\operatorname{Ext}^*_{kG}(M,N)$, so the latter is a finite module over $H(G)/\mathfrak{a}_G(M)$. It follows that there is an $n$ such that $\operatorname{Ext}^r_{kG}(M,N) = 0$ for $r \geq n$. In principle, $n$ could depend on $N$, but since there are only finitely many simple $kG$-modules, we can certainly choose a single $n$ to work for such $N$. Since any finitely generated $kG$-module $N$ has a filtration with simple factors, it follows that $\operatorname{Ext}^r_{kG}(M,N) = 0$ for $r \geq n$ for each such $N$. From this we may conclude in turn that $\operatorname{Ext}^1_{kG}(M,N) = 0$ for every finitely generated $kG$-module $N$, i.e. $M$ is $kG$-projective. For, consider the short exact sequence

$$0 \to C \to kG \otimes N \to kG \otimes_{kG} N = N \to 0.$$

Since

$$\operatorname{Ext}^n_{kG}(M, kG \otimes N) \cong \operatorname{Ext}^n_{kG}(M, \operatorname{Hom}(kG, N)) \cong \operatorname{Ext}^n_k(M,N) = 0,$$

for $n > 0$, we obtain

$$\operatorname{Ext}^n_{kG}(M,N) \cong \operatorname{Ext}^{n+1}_{kG}(M,C)$$

for $n \geq 1$. Hence, the vanishing of $\operatorname{Ext}^n_{kG}(M,-)$ for any positive $n$ implies its vanishing for $n = 1$. □

The following result of Chouinard (1976), which in some sense started the whole thing, is an immediate consequence of the above theorem and Theorem 8.3.1.

**Corollary 8.4.2** (Chouinard). *Let $G$ be a finite group, $k$ a field and $M$ a finitely generated $kG$-module. Then $M$ is projective if and only if its restriction to every elementary abelian p-subgroup is projective.*

Complexity may be defined somewhat more directly. Alperin's original approach ( (Alperin 1977) or (Alperin and Evens 1981)) was to consider the rate at which $\dim \operatorname{Ext}^n(M, N)$ grows. One way to relate this to what we did above is to consider the Poincaré function of a module. If $X$ is a graded vector space defined for non-negative degrees and finite dimensional in each degree, call

$$P(t) = \sum_n \dim_k X_n\, t^n$$

the Poincaré series for $X$. If $X$ is a finitely generated commutative graded $k$-algebra, then it is possible to show that the Poincaré series for $X$ is a rational function of the form

$$g(t)/(1-t^{j_1})(1-t^{j_2})\ldots(1-t^{j_d}) = h(t)/(1-t)^d$$

where $g(t) \in \mathbf{Z}[t]$ and $h(t)$ is a rational function in $\mathbf{Z}(t)$ with $h(1)$ defined and not 0. (The argument is a variant of the proof of the Hilbert Basis Theorem. See Benson (1984, Proposition 1.8.2).) If such is the case, then it is not hard to see that the function $f(n) = \dim_k X_n$ grows at the rate $n^{d-1}$ in that

$$f(n) < Cn^{d-1} \quad \text{for some constant } C$$

and the same is not true for any lesser power of $n$. In addition, one can show that $d$ is just the Krull dimension of the ring $X$. Here is a rough outline of the argument. If $d'$ is the Krull dimension of $X$, then, by the Noether Normalization Lemma (Atiyah and MacDonald 1969, Chapter 5, Exercise 16), we may choose a set of $d'$ homogeneous elements which generate a polynomial subring $R$ such that $X$ is a finite module over $R$. It follows by standard ring theory that $X$ and $R$ have the same Krull dimension. It is also not hard to see that they have the same growth rate. However, the Krull dimension of the polynomial ring is $d'$, and its Poincaré function has the form

$$1/(1-t^{j_1'})(1-t^{j_2'})\ldots(1-t^{j_{d'}'}) = h'(t)/(1-t)^{d'}$$

where $h'(1) \neq 0$.

Suppose now that $Y$ is a graded finitely generated $X$-module. It also has a Poincaré function which is rational as above but generally with a different $d$ and a different growth rate. $Y$ is a module over $X/\operatorname{ann}(Y)$ so it is not hard to see that the growth rate for $Y$ is bounded by the growth rate for $X/\operatorname{ann}(Y)$. (The growth rate of the middle term of a short exact sequence is the maximum of the growth rates of the other terms.) On the other hand, if $\{y_1, y_2, \ldots, y_r\}$ is a set of homogeneous generators of $Y$ as an $X$-module, then the homomorphism $X \to Y \oplus Y \oplus \cdots \oplus Y$ defined by

$$x \mapsto (xy_1, xy_2, \ldots, xy_r)$$

imbeds $X/\mathrm{ann}(Y)$ in a direct sum of copies of $Y$ and so the growth rate of $X/\mathrm{ann}(Y)$ is bounded by that of $Y$. Hence, the growth rate of $Y$ is the same as the growth rate of $X/\mathrm{ann}(Y)$ which is its Krull dimension.

We may now apply this reasoning to the case $X = H(G)$ and $Y = \mathrm{Ext}^*_{kG}(M, M)$ for $M$ a finitely generated $kG$-module. Thus, we see that $\dim_k \mathrm{Ext}^n_{kG}(M, M)$ grows at the rate $n^{d-1}$ where $d$ is the complexity of $M$. Similarly, if $N$ is any finitely generated $kG$-module, it follows that $\dim_k \mathrm{Ext}^n_{kG}(M, N)$ grows *at most* at the rate $n^{d-1}$.

**Theorem 8.4.3.** *If $M$ is a finitely generated $kG$-module of complexity $d$, then for every finitely generated $kG$-module $N$, $\dim_k \mathrm{Ext}^n_{kG}(M, N)$ grows at most at the rate $n^{d-1}$, and it grows at exactly that rate for $N = M$. Moreover, $\dim_k \mathrm{Ext}^n_{kG}(M, S)$ grows exactly at the rate $n^{d-1}$ for at least one simple $kG$-module $S$.*

**Proof.**

**Exercise 8.4.1.** Prove Theorem 8.4.3. □

The following is a special case of a result proved at the end of Section 10.1, but it can be proved directly by making use of an argument of Eisenbud (1980). (See also Alperin and Evens (1981, Section 3).)

**Proposition 8.4.4.** *Let $G$ be a finite group and $M$ a finitely generated $kG$-module. $cx_G(M) = 1$ if and only if $M$ is not projective, and $M$ has a periodic projective resolution.*

**Proof.** If $M$ is not projective and has a periodic projective resolution, then $\dim_k \mathrm{Ext}^n_{kG}(M, M)$ is bounded, so its growth rate is smaller than $n^0$. Thus, $cx_G(M) = d \leq 1$. Since $d = 0$ implies that $M$ is projective, we must have $d = 1$.

Conversely, suppose $M$ has complexity 1. If we decompose $M$ as a sum of indecomposable submodules, we can presume that $M = M' \oplus P$ where $P$ is projective and $M'$ has no non-trivial projective summands. It suffices to prove $M'$ has a periodic resolution, i.e. we may assume to start with that $M$ has no non-trivial projective summands.

By Proposition 8.1.1, for each finitely generated $N$, $\mathrm{Ext}^*_{kG}(M, N)$ is a finite module over $H(G)/\mathfrak{a}_G(M)$. That ring has dimension 1 by hypothesis, so, by the Noether Normalization Lemma (Atiyah and MacDonald 1969, Chapter 5, Exercise 16), each $\mathrm{Ext}^*_{kG}(M, N)$ is a finite module over a polynomial subring $k[\chi]$ of $H(G)$. By the structure theorem for finitely generated modules over a principal ideal domain, it follows that multiplication by $\chi$ is an isomorphism $\mathrm{Ext}^n_{kG}(M, N) \to \mathrm{Ext}^{n+j}_{kG}(M, N)$ for all $n > n_0$, where $j = \deg \chi$ and $n_0$ is chosen sufficiently large. $n_0$ depends in general on $N$, but if we restrict attention to the finite collection of *simple* $kG$-modules $N$, we can choose one $n_0$ which works for all.

## Complexity

Let $Y \to k$ be a projective resolution, and let $P \to M$ be the minimal projective resolution. Note that $Y \otimes P \to k \otimes M \cong M$ is also a projective resolution. Thus, we can choose a $G$ map of complexes

$$\Delta : P \to Y \otimes P.$$

It is is easy to check that the cup product pairing

$$H^*(G, k) \otimes \operatorname{Ext}^*_{kG}(M, N) \to \operatorname{Ext}^*_{kG}(M, N)$$

is induced from $\Delta$ through

$$\operatorname{Hom}_{kG}(Y, k) \otimes \operatorname{Hom}_{kG}(P, N) \to \operatorname{Hom}_{k(G \times G)}(Y \otimes P, N) \to \operatorname{Hom}_{kG}(P, N).$$

Let $f : Y_j \to k$ be a cocycle representing $\chi \in H^j(G, k)$, and let $\Phi = (f \otimes \operatorname{Id}) \circ \Delta : P \to P$. $\Phi$ is a map of complexes which *lowers* degree by $j$. Let $\Phi_n : P_{n+j} \to P_n$ denote the appropriate component of $\Phi$. Then one may check that the dual map

$$\Phi^n : \operatorname{Hom}_{kG}(P_n, N) \to \operatorname{Hom}_{kG}(P_{n+j}, N)$$

induces cup product by $\chi$ from $\operatorname{Ext}^n_{kG}(M, N)$ to $\operatorname{Ext}^{n+j}_{kG}(M, N)$. Since we have assumed $P \to M$ is minimal and $N$ is simple, it follows that the Homs are identical with the Exts, so $\Phi^n$ is an isomorphism for all simple $N$ and all $n$ sufficiently large. It follows readily from this that $\Phi_n : P_{n+j} \to P_n$ is an isomorphism for $n$ sufficiently large. Since the $\Phi_n$ form a map of complexes, it follows that the minimal resolution $P \to M$ is periodic from some point on. However, by Exercise 2.4.2, this implies that it is periodic in general since $M$ has no non-trivial projective summands. □

It is not too hard to see that a non-projective module $M$ without projective summands has a periodic projective resolution if and only if its minimal resolution is periodic. Such a module is called *periodic*.

# 9
# Stratification theorems

## 9.1 The Quillen stratification of $X_G$

Let $k$ be a field of characteristic $p$, $G$ a finite group, and $E$ an elementary abelian $p$-subgroup of $G$. We shall identify $E\hat{\phantom{x}} = \text{Hom}(E, \mathbf{F}_p)$ with $\text{Im}\{H^1(E, \mathbf{F}_p) \to H^1(E, k)\}$ for $p = 2$ or $\text{Im}\{H^2(E, \mathbf{Z}) \to H^2(E, k)\}$ for $p$ odd. (For $p$ odd, $E\hat{\phantom{x}}$ may also be thought of as the image of the composite map $H^1(E, \mathbf{F}_p) \xrightarrow{\delta} H^2(E, \mathbf{F}_p) \to H^2(E, k)$.) Similarly, we shall identify $E^* = \text{Hom}(E, k)$ with the $k$-subspace of $H(E)$ generated by $E\hat{\phantom{x}}$. (This is a subspace of degree 1 for $p = 2$ and degree 2 for $p > 2$.) For $p = 2$, $H(E) = H^*(E, k) \cong S(E^*)$ (the symmetric algebra on $E^*$). For $p > 2$, $H(E) = S(E^*) \oplus J$ where $J$ is the nilpotent ideal generated by $H^1(E, k)^2 \subset H^2(E, k)$; in fact, $J = \text{rad}\, H(E)$. In either case $H(E)/\text{rad}\, H(E) \cong S(E^*)$ is a domain, so $\mathfrak{p}_E = \text{res}_E^{-1}(\text{rad}\, H(E))$ is a prime ideal of $H(G)$. Also, $\mathfrak{p}_E$ is the radical of the ideal $\text{Ker}\,\text{res}_{E}$, and, as mentioned earlier, the closed subspace of $X_G$ which it defines is $\iota_E(X_E)$. One can view $\mathfrak{p}_E$ as a *generic point* of $\iota_E(X_E)$ in the sense that $\xi \in H(G)$ vanishes at $\mathfrak{p}_E$ (i.e. $\xi \in \mathfrak{p}_E$) if and only if it vanishes at every point $\mathfrak{p}$ of $\iota_E(X_E)$.

Since $G$ acts trivially on its own cohomology, it follows as in Proposition 8.2.3 that if $F$ is conjugate in $G$ to a subgroup of $E$, then $\mathfrak{p}_F \supseteq \mathfrak{p}_E$ so $\iota_F(X_F) \subseteq \iota_E(X_E)$. The following diagram clarifies these relations.

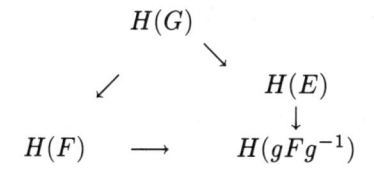

$$\begin{array}{ccc} & H(G) & \\ \swarrow & & \searrow \\ & & H(E) \\ & & \downarrow \\ H(F) & \longrightarrow & H(gFg^{-1}) \end{array}$$

Introduce the notation $F \leq_G E$ for '$F$ is conjugate in $G$ to a subgroup of $E$'. Similarly, use $F =_G E$ as the corresponding notion for '$F$ is conjugate to $E$'. The following result of Quillen (1971a, Theorem 2.7) will play an important role in what follows.

**Theorem 9.1.1.** *Let $E$ and $F$ be elementary abelian $p$-subgroups of $G$. Then $\mathfrak{p}_F \supseteq \mathfrak{p}_E$ (i.e. $\iota_F(X_F) \subseteq \iota_E(X_E)$) if and only if $F \leq_G E$. In particular, $\iota_F(X_F) = \iota_E(X_E)$ if and only if $F =_G E$.*

**Proof.** We shall construct an element $\zeta_F \in H(G)$ such that $\zeta_F \notin \mathfrak{p}_F$, but $\zeta_F \in \mathfrak{p}_E$ if $F$ is *not* conjugate to a subgroup of $E$. (That is, $\zeta_F$ does not vanish on $\iota_F(X_F)$ but does vanish on $\iota_E(X_E)$ for $F \not\leq_G E$.) Define

$$\epsilon_F = \prod_{0 \neq \beta \in \hat{F}} \beta.$$

$\epsilon_F$ has the property that it is invariant under any automorphism of $F$, and its restriction to any proper subgroup is trivial (since the restriction of at least one of its factors is trivial). Let

$$\mu_F = N_{F \to G}(1 + \epsilon_F).$$

Then, by the double coset formula, we have

$$\mathrm{res}_F(\mu_F) = \prod_{g \in D} N_{gFg^{-1} \cap F \to F}(\mathrm{res}_{gFg^{-1} \to gFg^{-1} \cap F}(1 + g^*\epsilon_F))$$

where $G = \bigcup_{g \in D} FgF$ is a double coset decomposition. Since $g^*$ carries $\hat{F}$ isomorphically onto $(gFg^{-1})\hat{\,}$,

$$g^*\epsilon_F = \epsilon_{gFg^{-1}},$$

and the only terms in the product not equal to 1 are those for which $gFg^{-1} = F$, i.e. $g \in N_G(F)$. Let $(N_G(F) : F) = qr$ with $q$ a power of $p$ and $(r, p) = 1$. It follows that $\mathrm{res}_F(\mu_F)$ is of the form

$$(1 + \epsilon_F)^{qr} = 1 + r\epsilon_F^q + \text{ higher degree terms}.$$

Let $\zeta_F$ be $r^{-1}$ times the homogeneous component of $\mu_F$ of the same degree as $\epsilon_F^q$. Then $\mathrm{res}_F(\zeta_F) = \epsilon_F^q$ which is not nilpotent. A similar calculation shows that if $F$ is not conjugate to a subgroup of $E$ then $\mathrm{res}_E(\mu_F) = 1$ and $\mathrm{res}_E(\zeta_F) = 0$. For in that case *all the terms are* 1. □

Note that the above argument shows that

$$\zeta_F \notin \mathfrak{p}_E \iff F \leq_G E \iff \mathfrak{p}_E \subseteq \mathfrak{p}_F. \tag{9.1}$$

For if $\mathfrak{p}_E \subseteq \mathfrak{p}_F$, then $\zeta_F \notin \mathfrak{p}_E$ which completes a circle of implications started in the proof of the theorem.

**Corollary 9.1.2.** *The minimal primes of $H(G)$ are the ideals $\mathfrak{p}_E$ for $E$ a maximal elementary abelian p-subgroup of $G$.*

**Proof.** We know by our basic decomposition of $X_G$ that

$$\mathrm{rad}(H(G)) = \bigcap_{\substack{E \text{ elem} \\ \text{abel } p\text{-sg}}} \mathfrak{p}_E.$$

Since any $\mathfrak{p}_F$ contains a $\mathfrak{p}_E$ with $E$ maximal, we may restrict attention to maximal $E$'s. What results is an irredundant primary decomposition of $\mathrm{rad}(H(G))$, so its constituents are in fact the minimal primes of $H(G)$. (See Atiyah and MacDonald (1969, Chapter 4).) □

Let $E$ be any elementary abelian $p$-group. Each subgroup $F$ of $E$ may be viewed as an $\mathbf{F}_p$-subspace. Define

$$X_E^+ = X_E - \bigcup_{F < E} \iota_{F \to E}(X_F).$$

Note that since every $F < E$ is a direct factor of $E$, it follows that $\mathrm{res}_{E \to F}$ is an epimorphism and $\iota_{F \to E}$ is a monomorphism onto a closed subspace. (In effect, $X_E$ may be thought of as an affine space, and $X_F$ as an affine subspace.) Since every $F < E$ is contained in a subspace of codimension 1, it follows that we may restrict the above union to $\mathbf{F}_p$-hyperplanes $F$. For such a hyperplane, *modulo radicals* we have

$$\mathrm{Ker}\,\mathrm{res}_{E \to F} = S(E^*)\eta_F$$

where $\eta_F \in \mathrm{Hom}(E, \mathbf{F}_p) \subseteq E^*$ is one of the $p-1$ forms with kernel $F$. It follows that $\iota_{F \to E}(X_F)$ is precisely the closed set in $X_E$ where $\eta_F$ vanishes. Hence, $X_E^+$ is the open set in $X_E$ where

$$\prod_{F \text{ a hyperplane}} \eta_F$$

does not vanish. On the other hand, if for each $F$, we replace $\eta_F$ by $\prod_{i=1}^{p-1} i\eta_F$, we do not change the open set. Hence, it follows that $X_E^+$ is the open subset of $X_E$ where

$$\epsilon_E = \prod_{\eta \neq 0} \eta$$

does not vanish. Moreover, by Atiyah and MacDonald (1969, Chapter 3, Problem 21), we know that the map $H(E) \to H(E)[\epsilon_E^{-1}]$ induces a homeomorphism of $|\mathrm{spec}\,H(E)[\epsilon_E^{-1}]|$ onto the open set $X_E^+ \subset X_E$.

Suppose now that $E$ is an elementary abelian $p$-subgroup of the finite group $G$. Define

$$X_{G,E}^+ = \iota_E(X_E) - \bigcup_{F < E} \iota_F(X_F)$$

where as above the union may also be taken over all $\mathbf{F}_p$-hyperplanes in $E$. Let $\zeta_E$ be the element in $H(G)$ defined in the proof of Theorem 9.1.1, and let $U$ be the subset of $\iota_E(X_E)$ on which $\zeta_E$ does not vanish. If $E$ is not conjugate to a subgroup of $F$, then we saw that $\zeta_E$ does vanish on $\iota_F(X_F)$ (i.e. $\zeta_E \in \mathfrak{p}_F$) so $U$ is disjoint from $\bigcup_{F<E} \iota_F(X_F)$. It follows that

$U \subseteq X_{G,E}^+$. On the other hand, $\operatorname{res}_E \zeta_E = \epsilon_E{}^q$ for some $p$-power $q$, so $\iota_E(X_E^+) \subseteq U$. (Think of $\epsilon_E{}^q$ as the 'composite' of $\zeta_E$ with $\iota_E$ and use the fact that $X_E^+$ is the subset of $X_E$ where $\epsilon_E$ does not vanish.) However, it is clear that $X_{G,E}^+ \subseteq \iota_E(X_E^+)$ so

$$X_{G,E}^+ = \iota_E(X_E^+) = U.$$

Note also that it follows from this that

$$\iota_E^{-1}(X_{G,E}^+) = X_E^+.$$

**Theorem 9.1.3** (Quillen Stratification, First Part). *Let $k$ be a field of characteristic $p$ and let $G$ be a finite group. Let $I$ be a family of elementary abelian $p$-subgroups of $G$, one from each conjugacy class. Then*

$$X_G = \bigcup_{E \in I} X_{G,E}^+$$

*is a decomposition of $X_G$ into disjoint locally closed affine subspaces.*

**Proof.** These sets are clearly disjoint, so it is clear from Corollary 8.3.3 and the above discussion that we obtain such a decomposition. $X_{G,E}^+$ is a locally closed subset of $X_G$, that is, it is the intersection of the closed subset $\iota_E(X_E)$ with the open set of all points at which $\zeta_E$ does not vanish. Finally,

$$X_{G,E}^+ \cong |\operatorname{spec} H(G)[\zeta_E^{-1}]/\mathfrak{p}_E[\zeta_E^{-1}]|$$

so $X_{G,E}^+$ is the underlying topological space of an affine scheme. □

To complete the discussion of Quillen Stratification, we need to examine the surjective map $X_E \to \iota_E(X_E) \subseteq X_G$ for $E$ an elementary abelian subgroup of $G$, and given the above discussion, we want to look in particular at $X_E^+ \to X_{G,E}^+$. In terms of the underlying rings, this amounts to understanding the imbeddings

$$\operatorname{Im} \operatorname{res}_{G \to E} \subseteq H(E)$$

and

$$(\operatorname{Im} \operatorname{res}_{G \to E})[(\epsilon_E{}^q)^{-1}] \subseteq H(E)[\epsilon_E{}^{-1}].$$

(Recall that $\operatorname{res} \zeta_E = \epsilon_E{}^q$ where $q$ is the $p$-power part of $(N_G(E) : E)$.) After factoring out radicals, we are led to ring extensions

$$R_{G \to E} = H(G)/\mathfrak{p}_E \subseteq S(E^*)$$

and

The Quillen stratification of $X_G$     113

$$R_{G \to E}[(\epsilon_E{}^q)^{-1}] \subseteq S(E^*)[\epsilon_E{}^{-1}].$$

Let $W = N_G(E)/C_G(E)$. $W$ acts faithfully on $E$ by conjugation, so it acts faithfully on $E^*$ and $S(E^*)$. Since it is possible to imbed $E$ in $X_E$ consistently with these actions, $W$ acts faithfully on $X_E$. (See the exercises.) Finally, since $X_E^+$ is dense in $X_E$, it also acts faithfully on $X_E^+$. Consider the towers of rings and subrings below:

$$\begin{array}{ccc} S(E^*) & \hookrightarrow & S(E^*)[\epsilon_E{}^{-1}] \\ \cup & & \cup \\ S(E^*)^W & \hookrightarrow & S(E^*)[\epsilon_E{}^{-1}]^W = S(E^*)^W[\epsilon_E{}^{-1}] \\ \cup & & \cup \\ R_{G \to E} & \hookrightarrow & R_{G \to E}[\epsilon_E{}^{-q}] \end{array}$$

All the rings in sight are noetherian rings and are finitely generated modules over the rings appearing below them.

**Lemma 9.1.4.** *For each element $\xi \in S(E^*)^W$, we have*

$$\xi^q \in R_{G \to E}[(\epsilon_E{}^q)^{-1}].$$

**Proof.** As in the proof of Theorem 9.1.1, consider

$$\Delta = N_{E \to G}(1 + \xi \epsilon_E) \in H(G, k).$$

Exactly as before, the double coset formula shows that

$$\operatorname{res}_E \Delta = 1 + r(\xi \epsilon_E)^q + \text{higher degree terms}$$

so that $\xi^q \epsilon_E{}^q \in \operatorname{Im} \operatorname{res}_E$. Factoring out the radicals yields

$$\xi^q \in R_{G \to E}[(\epsilon_E{}^q)^{-1}]$$

as claimed.  □

**Proposition 9.1.5.** *Let $k$ be a field of characteristic $p$, and let $A \subseteq B$ be $k$ algebras. Suppose there is a power $q$ of $p$ such that $\beta^q \in A$ for each $\beta \in B$. Then the induced map $|\operatorname{spec} B| \to |\operatorname{spec} A|$ is a bijection. In particular, for each prime ideal $\mathfrak{p}$ of $A$ there is a unique prime ideal $\mathfrak{P}$ of $B$ such that $A \cap \mathfrak{P} = \mathfrak{p}$.*

**Proof.**

**Exercise 9.1.1.** Hint: If $\mathfrak{p}$ is a prime ideal of $A$, show that $\{\beta \in B \mid \beta^q \in \mathfrak{p}\}$ is a prime ideal of $B$ contained in every prime ideal $\mathfrak{P}$ of $B$ such that $\mathfrak{P} \cap A = \mathfrak{p}$.  □

We apply this to the ring extension $R_{G \to E}[\epsilon_E^{-q}] \subseteq S(E^*)[\epsilon_E^{-1}]^W$. In our case, the rings are also noetherian, so, by Atiyah and MacDonald (1969, Chapter 6, Exercise 11), the corresponding map is *closed* and hence it is a homeomorphism. Thus, we may identify $X_{G,E}^+ = \iota_E(X_E^+) = |\operatorname{spec} R_{G \to E}[\epsilon_E^{-q}]|$ with $|\operatorname{spec} S(E^*)^W[\epsilon_E^{-1}]|$. On the other hand, since $S(E*)[\epsilon_E^{-1}]$ is a finite module over $S(E*)[\epsilon_E^{-1}]^W$, it follows that not only is

$$X_E^+ \to |\operatorname{spec} S(E^*)[\epsilon_E^{-1}]^W| = X_{G,E}^+$$

a surjection, but $W$ acts transitively on the pre-image of every point. (That is, $W$ transitively permutes the prime ideals in the over-ring above a given prime ideal in the subring. See Atiyah and MacDonald (1969, Chapter 5, Exercise 13).)

The above analysis gives us a rather complete picture of the relation between $X_G$ and the $X_E$ for $E$ elementary abelian. $X_G$ is covered disjointly by locally closed subsets $X_{G,E}^+$, one for each conjugacy class of elementary abelian $p$-subgroups, and each $X_{G,E}^+$ is the space of orbits of the corresponding $X_E^+$ under the action of the group $W = N_G(E)/C_G(E)$. We may also say this in slightly different language as follows. If $\mathfrak{p} \in X_G$ can be written $\iota_E(\mathfrak{q})$, where $E$ is an elementary abelian $p$-subgroup and $\mathfrak{q} \in X_E^+$, call $E$ a *vertex* and $\mathfrak{q}$ a *source* for $\mathfrak{p}$. (Thus $E$ is minimal with respect to the property that $\mathfrak{p} = \iota_E(\mathfrak{q})$ with $\mathfrak{q} \in X_E$.) (See Avrunin and Scott (1982) for the 'source' of this notion.) Then the following result is an easy consequence of the above analysis.

**Proposition 9.1.6.** *The source and vertex of an element in $X_G$ are unique up to conjugation by an element of $G$. More precisely, if $(E, \mathfrak{q})$ is a (vertex, source) for $\mathfrak{p}$ and $\mathfrak{p} = \iota_{E'}(\mathfrak{q}')$ then $\exists g \in G$ such that $gEg^{-1} \leq E'$ and $\iota_{E \to E'}(g^*(\mathfrak{q})) = \mathfrak{q}':$*

$$\begin{array}{ccc} & & G \\ & \nearrow & \\ E' & & \uparrow \\ \uparrow & & \\ gEg^{-1} & \to & E \end{array}$$

To complete this section we shall investigate further the action of $W$ on $X_E^+$. In particular, we shall show that in certain cases it acts *freely*.

Let $\Omega$ be a field extension of $k$, and consider

$$\Omega_E = \operatorname{Hom}_{k\text{-alg}}(H^*(E,k), \Omega) \cong \operatorname{Hom}_{k\text{-alg}}(S(E^*), \Omega)$$
$$\cong \operatorname{Hom}_k(E^*, \Omega) \cong E \otimes_{\mathbb{F}_p} \Omega.$$

The first isomorphism is a consequence of the fact that any algebra homomorphism is trivial on the ideal generated by elements of odd degree if $p$ is odd while for $p = 2$ the two sides are the same. The second isomorphism is obtained by restricting ring homomorphisms to $E^*$. The fact that it is an isomorphism follows from the universal mapping property defining the symmetric algebra. The isomorphism $E \otimes_{\mathbf{F}_p} \Omega \cong \mathrm{Hom}_k(E^*, \Omega)$ is obtained by sending $x \otimes a$ to $f$ where $f(\alpha) = a\alpha(x)$ for $\alpha \in E^*$. (The reader should try to define an inverse. Choosing an $\mathbf{F}_p$-basis for $E$ could help.)

If $A$ is a commutative $k$-algebra with $X = |\mathrm{spec}\, A|$, we shall call $\mathrm{Hom}_{k\text{-alg}}(A, \Omega)$ the set of *points* of $X$ in $\Omega$. This is an abuse of terminology since the points in a given $\Omega$ depend on $\mathrm{spec}\, A$, i.e. on the ring $A$, not on the topological space $X$ alone. For $A = S(E^*)$, $\Omega_E$ denotes the set of points of $X_E$ in $\Omega$.

The two most interesting cases to consider are $\Omega$ the algebraic closure of $k$ and $\Omega$ an algebraically closed extension of $k$ of sufficiently high transcendence degree. The first case is sufficient to extract the maximal ideals of $H(E)$ (or equivalently $S(E^*)$), while the second case allows us to extract the prime ideals.

Returning to our analysis, we see that $\mathrm{Hom}_{k\text{-alg}}(S(E^*)[\epsilon_E^{-1}], \Omega)$ can be identified with the subset of $\mathrm{Hom}_{k\text{-alg}}(S(E^*), \Omega)$ of morphisms which do not vanish on $\epsilon_E$ (and so do not vanish on any $\mathbf{F}_p$-linear form). Hence, under the above isomorphisms, $\mathrm{Hom}_{k\text{-alg}}(S(E^*)[\epsilon_E^{-1}], \Omega)$ (which is the set of points of $X_E^+$ in $\Omega$) corresponds to $(E \otimes_{\mathbf{F}_p} \Omega)^+$, the complement in $E \otimes_{\mathbf{F}_p} \Omega$ of the union of the hyperplanes defined by linear forms with coefficients in $\mathbf{F}_p$.

It is easy to check that these isomorphisms are $GL(E)$-isomorphisms, so to determine the action of $W$ on the set of points in $\Omega$, it suffices to investigate its action on $E \otimes_{\mathbf{F}_p} \Omega$. Let $w \neq 1 \in W$. Then

$$(E \otimes_{\mathbf{F}_p} \Omega)^w = E^w \otimes_{\mathbf{F}_p} \Omega.$$

(The reader should prove this. Choosing an $\mathbf{F}_p$-basis for $\Omega$ could help.) Hence, if $w$ fixes a non-zero element of $E \otimes_{\mathbf{F}_p} \Omega$, that element must be in a subspace with a basis from $E = E \otimes_{\mathbf{F}_p} \mathbf{F}_p$, so it is contained in the kernel of a linear form $\alpha$ with coefficients in $\mathbf{F}_p$, i.e. in a hyperplane defined over $\mathbf{F}_p$.

**Theorem 9.1.7** (Quillen Stratification, Second Part). *If $E$ is an elementary abelian $p$-subgroup of $G$, then $W = N_G(E)/C_G(E)$ acts freely on the points of $X_E^+$ in $\Omega$ for each field extension $\Omega$ of $k$. In particular, if $k$ is algebraically closed, then $W$ acts freely on the set of closed points (maximal ideals) of $X_E^+$.*

**Proof.** The first part has been proved. For the second part, define a map $\Omega_E \to X_E$ by associating with each $k$-algebra homomorphism $h: S(E^*) \to \Omega$ its kernel $\mathfrak{p}$ which is a prime ideal of $H(E)$ (or equivalently of $S(E^*)$). It is easy to see that this map is consistent with the action of $W$. If we take $\Omega$ to be the algebraic closure of $k$, then, by the Hilbert *Nullstellensatz*, this map is onto the set of of maximal ideals, i.e. the set of closed points in $X_E$. (The reader unfamiliar with the relevant commutative algebra should review this material, say in Atiyah and MacDonald (1969, Chapter 5, Exercises 17–21) or Bourbaki (1972).) If $k$ is algebraically closed ($k = \Omega$), the map is one-to-one, since in general $S(E^*)/\mathfrak{m} \cong \operatorname{Im} h$ is isomorphic to an algebraic extension of $k$ where the isomorphism preserves $k$. □

**Exercise 9.1.2.** Let $G$ be the dihedral group of order 8, and let $k = \mathbf{F}_2$. Let $E$ be the normal elementary abelian subgroup of order 4. Then $W = \{1, z\}$ is cyclic of order 2. We may assume $S(E^*) = \mathbf{F}_2[\chi, \eta]$ where $z\chi = \chi + \eta$, $z\eta = \eta$, and $\epsilon_E = \chi\eta(\chi + \eta)$.

(a) Show that the ideal $\mathfrak{m}$ in $\mathbf{F}_2[\chi, \eta][\epsilon_E^{-1}]$ generated by $\chi^2 + \chi + 1$ and $\eta + 1$ is maximal and fixed by $z$.

(b) Show that if we tensor with $L$, the quadratic extension of $\mathbf{F}_2$, then $\mathfrak{m}$ splits into two maximal ideals which are permuted by $z$.

One consequence of the above theorem is that

$$R' = S(E^*)[\epsilon_E^{-1}]^W \subseteq S' = S(E^*)[\epsilon_E^{-1}]$$

is a a Galois extension in the sense of Chase, Harrison, and Rosenberg (1964). It follows that $S'$ is a separable extension of $R'$, and much of classical Galois theory becomes available. (See also Priddy and Wilkerson (1985).) In particular, the 'normal basis theorem' holds in the sense that $S'$ is a projective $R'W$-module. (In the classical case, it is free of rank 1.) Thus, although $W$ may not act freely on the points of $\operatorname{spec} S'$, it is still proper to think of $\operatorname{spec} S'$ as an unramified covering of $\operatorname{spec} R'$ with covering group $W$.

**Exercise 9.1.3.** To establish that $R' \subseteq S'$ is a Galois extension, verify criterion (f) of Theorem 1.3 (Chase, *et al.* 1964): for each $w \in W$, and each maximal ideal $\mathfrak{M}$ of $S'$, there is an $s \in S'$ such that $w(s) - s \notin \mathfrak{M}$. Hint: Choose an algebra homomorphism $\phi : S' \to \Omega$ with kernel $\mathfrak{M}$ (where $\Omega$ is the algebraic closure of $k$). Use Theorem 9.1.7.

There is one result in the Galois theory of rings concerning the *trace* $\operatorname{Tr}_W = \sum_{w \in W} w : S' \to R'$, which we shall need later, so we provide a proof here. (See Chase, *et al.* (1964, Lemma 1.6) which attributes the result to an earlier paper of Auslander and Goldman.)

**Corollary 9.1.8.** *With the above notation, $1 \in S(E^*)[\epsilon_E^{-1}]^W$ is the trace of some element in $S(E^*)[\epsilon_E^{-1}]$. Equivalently, some power of $\epsilon_E$ is the trace of an element $\mu \in S(E^*)$.*

**Proof.** It suffices to prove the corollary in the case where $k$ is algebraically closed. To see this, note first that if $L$ is a field extension of $k$, then

$$S_L(\operatorname{Hom}_{\mathbf{F}_p}(E, L)) \cong S_L(\operatorname{Hom}_{\mathbf{F}_p}(E, k) \otimes_k L) \cong S_k(\operatorname{Hom}_{\mathbf{F}_p}(E, k)) \otimes_k L$$

(where we let $S_k(V)$ denote the symmetric algebra over $k$ of the vector space $V$ defined over $k$). Let $A = S_k(\text{Hom}_{\mathbf{F}_p}(E, k))$. It is not hard to see that $(A \otimes_k L)[(\epsilon_E \otimes 1)^{-1}] \cong A[\epsilon_E^{-1}] \otimes_k L$. Let $\tilde{A} = A[\epsilon_E^{-1}]$. We want to show that if $1 \in (\tilde{A} \otimes_k L)^W = \tilde{A}^W \otimes_k L$ is the trace of some element in $\tilde{A} \otimes_k L$, then it is already the trace of some element in $\tilde{A}$. However, any $k$-splitting $L = k \oplus L'$ yields a $kW$-splitting $\tilde{A} \otimes_k L = \tilde{A} \oplus \tilde{A} \otimes_k L'$, and similarly for the $W$-invariants. It follows that any element of $\tilde{A}^W$ which is the trace of something in the sum is already the trace of something in $\tilde{A}$ as claimed.

Suppose now that $k$ is algebraically closed. We shall show that

$$\text{Im Tr}_W = S(E^*)[\epsilon_E^{-1}]^W.$$

If not, there is a maximal ideal $\mathfrak{m}$ in that ring which contains $\text{Im Tr}_W$. Let $\mathfrak{M}$ be a maximal ideal of $S(E^*)[\epsilon_E^{-1}]$ lying over $\mathfrak{m}$. By Theorem 9.1.7, the ideals $w(\mathfrak{M}), w \in W$ are distinct. By the Chinese Remainder Theorem, we may choose $\xi \in S(E^*)[\epsilon_E^{-1}]$ such that $\xi \equiv 1 \bmod \mathfrak{M}$ and $\xi \equiv 0 \bmod w(\mathfrak{M})$ for $w \neq 1$. Then, $\text{Tr}_W \xi \equiv 1 \bmod \mathfrak{M}$, and this contradicts $\text{Tr}\,\xi \in \mathfrak{m}$. □

**Exercise 9.1.4.** Let $E$ be an elementary abelian $p$-group. We shall view $X_E$ as $|\text{spec}\,S(E^*)|$. Define a map $j : E \to X_E$ by taking $j(x)$ to be the ideal generated by all $\alpha \in S(E^*)$ which vanish at $x$.
 (a) Show that $j$ is one-to-one.
 Let $\phi$ be an automorphism of $E$.
 (b) Show that $j$ is consistent with the induced action of $\phi$ on $X_E$.
 (c) Show that $\phi$ acts continuously on $X_E$.

## 9.2 Quillen's homeomorphism

The product of the restriction maps

$$\text{res} : H(G) \to \prod_{\substack{E \text{ elem abel} \\ p-\text{subgp}}} H(E)$$

maps $H(G)$ into the subring $I_G$ of elements whose components are coherently related by restriction maps and conjugations $x^*$ by elements of $G$. We showed in Corollary 8.3.4 that the kernel of this map is nilpotent. Moreover, it is easy to see that $I_G$ is a finite module over $H(G)$. (In fact, $\prod_E H(E)$ is finite over $H(G)$ since each $H(E)$ is. See Corollary 7.4.7.) It follows that the induced map

$$|\text{spec}\,I_G| \to |\text{spec}(\text{Im res})| = |\text{spec}\,H(G)| \qquad (9.2)$$

is onto. In this section, we shall show that the cokernel of map $H(G) \to I_G$ is also 'nilpotent' in the sense that there is a power $q$ of $p$ such that

$$\rho^q \in \operatorname{Im} \operatorname{res} \qquad \text{for all } \rho \in I_G.$$

It follows from this that the map (9.2) is a homeomorphism. (However, the $p$-power property is generally stronger. See below.)

The analysis in this section is inspired in part by Quillen (1971c, Appendix B).

The ring $I_G$ is the projective (inverse) limit of the following functor. Let $\mathcal{E}$ denote the category whose objects are the elementary abelian subgroups of $G$ and whose maps are inclusions of one such subgroup in another followed by conjugations $c_x$ by elements $x \in G$. Consider the functor given by specifying

$$E \mapsto H(E)$$
$$E \leq F \mapsto \operatorname{res}: H(F) \to H(E)$$
$$E \xrightarrow{c_x} xEx^{-1} \mapsto c_x^*: H(xEx^{-1}) \to H(E).$$

Then, $I_G = \varprojlim H(E)$.

If $B$ is a subset of a ring, let $B^{[q]}$ denote the set of $q$th powers of elements of $B$ to distinguish it from $B^q$ which is the additive subgroup generated by all products of $q$ elements of $B$. We noted in Proposition 9.1.5 that if $A \supseteq B$ is an extension of $k$-algebras with $B^{[q]} \subseteq A$ for some $p$-power $q$, then $|\operatorname{spec} B| \to |\operatorname{spec} A|$ is a homeomorphism. The converse of Proposition 9.1.5 is not true without further assumptions. For example, let $A$ be a field and let $B$ be a finite extension of $A$ which is not purely inseparable. Then, both $|\operatorname{spec} B|$ and $|\operatorname{spec} A|$ reduce to a single point and the map between them is a homeomorphism, but $B^{[q]} \not\subseteq A$ for any $p$-power $q$. However, we have the following partial converse.

**Theorem 9.2.1.** *Let $k$ be a field of characteristic $p$, and let $A \subseteq B$ be $k$-algebras with $A$ noetherian and $B$ be a finite module over $A$. Suppose that for each prime ideal $\mathfrak{p}$ in $A$ there is a unique prime ideal $\mathfrak{P}$ in $B$ such that $\mathfrak{P} \cap A = \mathfrak{p}$. Suppose moreover that, for each such $\mathfrak{p}$, $(B/\mathfrak{P})^{[q]}$ is contained in the field of fractions of $A/\mathfrak{p}$ for some $p$-power $q$. Then $B^{[q]} \subseteq A$ for some $p$-power $q$.*

Note that the hypothesis that the map on prime ideals is one-to-one (uniqueness of $\mathfrak{P}$ given $\mathfrak{p}$) in the above circumstances is equivalent to asserting that $|\operatorname{spec} B| \to |\operatorname{spec} A|$ is a homeomorphism. The hypothesis on the quotient rings says that the field of fractions of $B/\mathfrak{P}$ is a purely inseparable extension of the field of fractions of $A/\mathfrak{p}$.

**Proof.** Consider the ideals in $A$

$$(A:B;d) = \{\alpha \in A \,|\, \alpha B^{[p^d]} \subseteq A\}.$$

It is easy to check that they form an ascending chain, so, since $A$ is noetherian, there is a $d$ such that $(A:B;d') = (A:B;d)$ for all $d' \geq d$. If the conclusion of the theorem is false, $(A:B;d)$ is a proper ideal of $A$, so we may choose a minimal prime ideal $\mathfrak{p}$ of $A$ containing it. (See Atiyah and MacDonald (1969, Proposition 4.6 and Theorem 7.13).) Let $S = A - \mathfrak{p}$, $A_\mathfrak{p} = A[S^{-1}]$, and $B_\mathfrak{p} = B[S^{-1}]$. Then, it is clear that $(A:B;d')[S^{-1}] = (A:B;d)[S^{-1}]$ for all $d' \geq d$. In addition, for any $d'$, $(A_\mathfrak{p} : B_\mathfrak{p}; d') = (A:B;d')[S^{-1}]$. (We leave the proof to the reader as an exercise in localization. It uses the fact that $B$ is a finite module over $A$.) Moreover, $(A:B;d)[S^{-1}]$ is a proper ideal of $A_\mathfrak{p}$. For, if $1 = \alpha/\sigma$ with $\alpha \in (A:B;d)$ and $\sigma \in S$, there must be a $\tau \in S$ such that $\tau(\sigma - \alpha) = 0$ or $\tau\sigma = \tau\alpha \in \mathfrak{p}$, which is impossible.

Because of the minimality of $\mathfrak{p}$, the only prime ideal of $A_\mathfrak{p}$ containing $(A:B;d)[S^{-1}]$ is the maximal ideal $\mathfrak{p}[S^{-1}]$. Hence, the latter is the radical of the former. Moreover, $B_\mathfrak{p}$ certainly is a finite module over $A_\mathfrak{p}$, and it is easy to see that the ring extension $A_\mathfrak{p} \subseteq B_\mathfrak{p}$ inherits the hypotheses of the theorem. In effect, these remarks reduce the proof to the case $A$ is a local ring with maximal ideal $\mathfrak{p}$. (For, if $B_\mathfrak{p}^{[p^d]} \subseteq A_\mathfrak{p}$ for at least one $d$, $(A_\mathfrak{p} : B_\mathfrak{p}; d')$ cannot be a proper ideal for all sufficiently large $d'$.)

We now assume that $A$ is a local ring and the radical of $(A:B;d)$ is the unique maximal ideal $\mathfrak{p}$. Let $\mathfrak{P}$ be the unique prime ideal of $B$ such that $\mathfrak{P} \cap A = \mathfrak{p}$. By the 'Going-up Theorems' (Atiyah and MacDonald 1969, Corollary 5.8 ff), $\mathfrak{P}$ is the only maximal ideal of $B$, so $B$ is also a local ring. Moreover, $\mathfrak{P}$ is the radical of $B\mathfrak{p}$, since any prime ideal of $B$ containing $B\mathfrak{p}$ must contract to $\mathfrak{p}$. Hence, there is a $p$-power $q_1$ such that $\mathfrak{P}^{[q_1]} \subseteq B\mathfrak{p}$. Similarly, since $\mathfrak{p}$ is the radical of $(A:B;d)$, $\mathfrak{p}^{[q_2]} \subseteq (A:B;d)$ for some $p$-power $q_2$, i.e. $\mathfrak{p}^{[q_2]} B^{[p^d]} \subseteq A$. Let $q_3 = \max(q_2, p^d)$. Then $(B\mathfrak{p})^{[q_3]} = \mathfrak{p}^{[q_3]} B^{[q_3]} \subseteq \mathfrak{p}^{[q_2]} B^{[p^d]} \subseteq A$. Hence, $\mathfrak{P}^{[q_1 q_3]} \subseteq A$. On the other hand, by hypothesis, $B/\mathfrak{P}$ is a purely inseparable extension of $A/\mathfrak{p}$, so $B^{[q_4]} \subseteq A + \mathfrak{P}$ for some $p$-power $q_4$. Putting this all together yields $B^{[q_4 q_1 q_3]} \subseteq A^{[q_1 q_3]} + \mathfrak{P}^{[q_1 q_3]} \subseteq A$ as required. $\square$

We want to apply the previous result to the case $B = I_G$ and $A = \operatorname{Im}\operatorname{res} : H(G) \to I_G$. Our strategy will be to push the stratification results of the previous section to the ring $I_G$. Since $B = I_G$ is a projective limit, we have a projection $p_E : B \to H(E)$ for each elementary abelian $p$-subgroup $E$. Let $\mathfrak{P}_E = p_E^{-1}(\operatorname{rad} H(E))$. Clearly, $\mathfrak{P}_E \supseteq \mathfrak{P}_{E'}$ if $E \leq_G E'$ with equality if $E =_G E'$. Let $\zeta'_E$ denote the image in $A$ of the element $\zeta_E$ defined in the proof of Theorem 9.1.1. We have $p_E(\zeta'_E) = \operatorname{res}_{G \to E}(\zeta_E) = \epsilon_E{}^q$ for some $p$-power $q$. Thus, since $\mathfrak{p}_E$ is the pull back to $H(G)$ of $\mathfrak{P}_E$, it follows from (9.1) that

$$\zeta'_F \notin \mathfrak{P}_E \cap A \iff F \leq_G E \iff \mathfrak{P}_E \subseteq \mathfrak{P}_F.$$

Hence, the ideals $\mathfrak{P}_E$ associated with distinct conjugacy classes of elementary abelian $p$-subgroups $E$ are distinct.

For each elementary abelian $p$-subgroup $E$, consider the tower

$$\frac{H(G)}{p_E}[\epsilon_E{}^{-q}] \subseteq \frac{B}{\mathfrak{P}_E}[\epsilon_E{}^{-q}] \subseteq \mathcal{S}(E^*)^W[\epsilon_E{}^{-q}]$$

where we have made appropriate innocuous identifications. We showed in Lemma 9.1.4 that the $[q]$th power of the top level is contained in the bottom level, so the same is true of the intermediate level.

Suppose that $\mathfrak{Q}$ is a prime ideal of $B$. Since $\prod_E H(E)$ is an integral extension of $B$ (even of $A$), it follows that $\mathfrak{Q}$ is the contraction of some prime ideal of $\prod_E H(E)$. It follows from this that $\mathfrak{Q} = p_E^{-1}(\mathfrak{q}')$ for some prime ideal $\mathfrak{q}'$ in $H(E)$ for some $E$. In particular, $\mathfrak{Q} \supseteq \mathfrak{P}_E$. Choose a minimal such $E$; it follows that $\epsilon_E \notin \mathfrak{q}'$. Then $\zeta'_E \notin \mathfrak{Q} \cap A$.

Choose representatives $E_i$ of the distinct conjugacy classes of elementary abelian $p$-subgroups of $G$, let $\mathfrak{P}_i = \mathfrak{P}_{E_i}$, and let $\zeta_i = \zeta'_{E_i} \in A$. Then we have verified the following assertions.

(a) $\zeta_i \notin \mathfrak{P}_j \cap A$ if and only if $\mathfrak{P}_j \subseteq \mathfrak{P}_i$.
(b) For each prime ideal $\mathfrak{Q}$ of $B$, $\exists\ i$ such that $\mathfrak{P}_i \subseteq \mathfrak{Q}$ and $\zeta_i \notin \mathfrak{Q} \cap A$.
(c) For each $i$, there is a power $q$ of $p$ such that

$$(B/\mathfrak{P}_i)[\overline{\zeta}_i{}^{-1}]^{[q]} \subseteq (A/\mathfrak{p}_i)[\overline{\zeta}_i{}^{-1}]$$

where $\mathfrak{p}_i = A \cap \mathfrak{P}_i$.

**Lemma 9.2.2.** *Under the above circumstances, for each prime ideal $\mathfrak{q}$ in $A$, there is a unique prime ideal $\mathfrak{Q}$ in $B$ such that $A \cap \mathfrak{Q} = \mathfrak{q}$. Moreover, for each such $\mathfrak{q}$, there is a $p$-power $q$ such that $(B/\mathfrak{Q})^{[q]}$ is contained in the field of fractions of $A/\mathfrak{q}$.*

**Proof.** Since $B$ is integral over $A$, it follows that there exists a prime ideal $\mathfrak{Q}$ of $B$ such that $A \cap \mathfrak{Q} = \mathfrak{q}$. Suppose $A \cap \mathfrak{Q} = A \cap \mathfrak{Q}' = \mathfrak{q}$. By (b), choose $i$ such that $\mathfrak{Q} \supseteq \mathfrak{P}_i$ and $\zeta_i \notin \mathfrak{Q} \cap A = \mathfrak{q}$, and similarly choose $j$ such that $\mathfrak{Q}' \supseteq \mathfrak{P}_j$ and $\zeta_j \notin \mathfrak{q}$. However, $\zeta_i \notin \mathfrak{P}_j \cap A$ since the latter is a subset of $\mathfrak{Q}' \cap A = \mathfrak{q}$, so by (a), $\mathfrak{P}_j \subseteq \mathfrak{P}_i$. Reversing the roles of $i$ and $j$ shows that $\mathfrak{P}_i = \mathfrak{P}_j$, i.e. $i = j$. Consider next the ideals $\overline{\mathfrak{Q}} = \mathfrak{Q}/\mathfrak{P}_i$ and $\overline{\mathfrak{Q}}'$ in $\overline{B} = B/\mathfrak{P}_i$. These have the same contractions $\overline{\mathfrak{q}}$ to $\overline{A} = A/\mathfrak{p}_i$. Hence, after localizing with respect to the powers of $\overline{\zeta}_i$, the same is true of the localized ideals. However, the localized rings satisfy the hypotheses of Proposition 9.1.5, so the localized ideals are the same; hence $\overline{\mathfrak{Q}} = \overline{\mathfrak{Q}'}$. Since $\mathfrak{Q}, \mathfrak{Q}' \supseteq \mathfrak{P}_i$, it follows that they are equal.

The assertion about residue ring extensions is clear since it is already true for
$$(A/\mathfrak{q})[\overline{\zeta}_i^{-1}] \subseteq (B/\mathfrak{Q})[\overline{\zeta}_i^{-1}].$$  □

Note that the first conclusion of the theorem amounts to the assertion that
$$|\operatorname{spec} I_G| \to |\operatorname{spec} \operatorname{Im} \operatorname{res}| = |\operatorname{spec} H(G)|$$
is a homeomorphism. In addition, the conclusions of the theorem are the hypotheses of Theorem 9.2.1, so we obtain the result promised at the beginning of this section.

**Corollary 9.2.3.** *Let $G$ be a finite group and $k$ a field of characteristic $p$. Let $I_G$ denote the subring of $\prod_E H(E)$ of coherently related elements. There is a pth power $q$ such that*
$$I_G{}^{[q]} \subseteq \operatorname{Im}\{H(G) \to I_G\}.$$

Consider the functor from the category $\mathcal{E}$ of elementary abelian $p$-subgroups of $G$ to the category of topological spaces which attaches to each object $E$ in $\mathcal{E}$ the space $X_E$ and to each morphism in $\mathcal{E}$ the induced continuous map. Consider the inductive (direct) limit $\varinjlim X_E$ of this functor. (It may be thought of as a collection of affine spaces suitably glued together along subspaces rational over $\mathbf{F}_p$ with additional identifications arising from the actions of the groups $W$.) While it is not true in general that the functor $A \mapsto |\operatorname{spec} A|$ takes projective limits to inductive limits, in this case we do have

$$|\operatorname{spec} I_G| \cong \varinjlim_{\substack{E \leq G \\ E \text{ elem}}} X_E. \tag{9.3}$$

The easiest way to see this is to prove first that the natural map
$$X_G \to \varinjlim X_E$$
is a homeomorphism by using the results in Section 9.1. Since the map (9.3) is a factor of this map, it follows from the above discussion that (9.3) is also a homeomorphism.

**Exercise 9.2.1.** Let $k$ be a field of characteristic $p$. A homomorphism $f: A \to B$ of $k$-algebras is called an $F$-*isomorphism* if its kernel is nilpotent and $B^{[q]} \subseteq \operatorname{Im} f$ for some $p$-power $q$.

Let $f: A \to B$ be a homomorphism of $k$-algebras with $A$ noetherian and $B$ a finite module over $A$.

Show that the induced map $\operatorname{Hom}_{k\text{-alg}}(A, \Omega) \to \operatorname{Hom}_{k\text{-alg}}(B, \Omega)$ is a homeomorphism (for the Zariski topologies) for every algebraically closed field $\Omega$ of characteristic $p$ if and only if $f$ is an $F$-isomorphism. Hint: Use Theorem 9.2.1. (See Quillen (1971c, Propositions B.8 and B.9).)

## 9.3  Avrunin–Scott stratification

Avrunin and Scott (1982) extend the Quillen Stratification Theorem by showing that $X_G(M)$ is the union of the disjoint subsets

$$X_{G,E}(M)^+ = \iota_E(X_E(M) \cap X_E^+).$$

We devote this section to this result.

The start of the proof is straightforward. Let $\mathfrak{p} \in X_G(M)$, and let $(E, \mathfrak{q})$ be a (vertex, source) for $\mathfrak{p}$. By the theorem of Alperin–Avrunin–Evens (Theorem 8.3.1),

$$X_G(M) = \bigcup_{E \text{ elem}} \iota_E(X_E(M)),$$

so we can choose $E'$ and $\mathfrak{q}' \in X_{E'}(M)$ such that $\iota_{E'}(\mathfrak{q}') = \mathfrak{p}$. By Proposition 8.2.3, we may assume that $E \leq E'$ and $\iota_{E \to E'}(\mathfrak{q}) = \mathfrak{q}'$. Thus, $\mathfrak{q} \in \iota_{E \to E'}^{-1}(X_{E'}(M))$, so it would suffice to prove the following result.

**Lemma 9.3.1** (Subgroup Lemma). *If $E \leq F$ are elementary abelian $p$-subgroups and $M$ is a $kF$-module, then $\iota_{E \to F}^{-1}(X_F(M)) = X_E(M)$.*

The assertion holds in general for a group $G$ and subgroup $H$. That result follows immediately as a corollary to the special case for elementary abelian $p$-groups (see below). The original proof of the Subgroup Lemma was quite involved and made use of Carlson's 'shifted subgroups' or equivalent arguments using restricted enveloping algebras of $p$-Lie algebras. (See Benson (1984, Sections 2.25, 2.26) for a treatment using shifted subgroups or Avrunin and Scott (1982) for one using $p$-Lie algebras.) The much simpler proof below is adapted from one invented by Stefan Jackowski. (See Evens and Jackowski (1990).)

**Proof.** We proceed by induction on $(F : E)$. Suppose first that $(F : E) = p$ so that $F = E \times P$ where $P = F/E$ is cyclic of order $p$. Note that because of the splitting, $\text{res}_{F \to E}$ is onto ($\text{res}_{F \to E} \inf_{E \to F} = \text{Id}$). We shall use this fact below.

We know that $\iota_E(X_E(M)) \subseteq X_F(M)$, so $X_E(M) \subseteq \iota_E^{-1}(X_F(M))$. Hence, it suffices to show that $X_E(M) \supseteq \iota_E^{-1}(X_F(M))$. The right hand side is the set of $\mathfrak{p} \in X_E$ such that $\mathfrak{P} = \iota_E(\mathfrak{p}) \supseteq \mathfrak{a}_F(M)$, so it suffices to show that

$$\mathfrak{p} \not\supseteq \mathfrak{a}_E(M) \Rightarrow \mathfrak{P} \not\supseteq \mathfrak{a}_F(M).$$

Suppose there is an element $\zeta \in \mathfrak{a}_E(M)$ such that $\zeta \notin \mathfrak{p}$. We shall find an element $\rho \in \mathfrak{a}_F(M)$ such that $\text{res}_{F \to E}(\rho) = \zeta^n$ for some $n$. It is clear for such an element that $\rho \notin \mathfrak{P} = \text{res}_{F \to E}^{-1}(\mathfrak{p})$.

Choose $\chi \in H(F)$ such that $\operatorname{res}_{F \to E}(\chi) = \zeta$. Consider the spectral sequence
$$H^*(P, \operatorname{Ext}^*_{kE}(M, M)) \Rightarrow \operatorname{Ext}^*_{kF}(M, M)$$
as before, as a module over the spectral sequence
$$H^*(P, H^*(E, k)) \Rightarrow H^*(F, k).$$
(The latter spectral sequence in fact collapses totally in the sense that the limit is naturally isomorphic to the $E_2$-term as a ring.) The action of $\chi \in H(F)$ on
$$E_\infty^{r,*} = F^r \operatorname{Ext}^*_{kF}(M, M) / F^{r+1} \operatorname{Ext}^*_{kF}(M, M)$$
arises from the action of $\zeta = \operatorname{res}_{F \to E} \chi$ on $E_2^{r,*} = H^r(P, \operatorname{Ext}^*_{kE}(M, M))$ which is trivial since by assumption $\zeta \operatorname{Ext}^*_{kE}(M, M) = 0$. It follows by taking $r = 0$ that
$$\chi \operatorname{Ext}^*_{kF}(M, M) \subseteq F^1 \operatorname{Ext}^*_{kF}(M, M),$$
and by taking $r = 1$ that
$$\chi^2 \operatorname{Ext}^*_{kF}(M, M) \subseteq F^2 \operatorname{Ext}^*_{kF}(M, M).$$
By Quillen–Venkov (Lemma 8.3.2), $F^2 \operatorname{Ext}^*_{kF}(M, M) = \beta_E \operatorname{Ext}^*_{kF}(M, M)$ where $\beta = \beta_E \in \operatorname{Ker} \operatorname{res}_{F \to E}$. Let $\xi = \chi^2$ and choose module generators $\mu_1, \ldots, \mu_m$ for $\operatorname{Ext}^*_{kF}(M, M)$ over $H(F)$. Then we have
$$\xi \mu_i = \beta \sum_{j=1}^m \alpha_{ij} \mu_j = \sum_j \beta \alpha_{ij} \mu_j \qquad i = 1, \ldots, m$$
where $\alpha_{ij} \in H(F)$. Put $\gamma_{ij} = \beta \alpha_{ij}$ and note that $\gamma_{ij} \in \operatorname{Ker} \operatorname{res}_{F \to E}$. Let $C$ denote the square matrix $(\gamma_{ij})$, and let $f(T) = \det(TI - C)$ be its characteristic polynomial. By the Cayley–Hamilton Theorem we have
$$f(\xi) \mu_i = 0 \qquad \text{for } i = 1, \ldots, m$$
so that $\rho = f(\xi) \in \mathfrak{a}_F(M)$. However, $f(\xi)$ restricts to $\det(\zeta^2 I - 0) = \zeta^{2m}$ in $H(E)$ so we have constructed the desired $\rho$.

Suppose now that $(F : E) > p$ and choose $E'$ such that $F > E' > E$. Then applying induction to the intermediate levels shows
$$\iota_{E \to F}{}^{-1}(X_F(M)) = \iota_{E \to E'}{}^{-1}(\iota_{E' \to F}{}^{-1}(X_F(M)))$$
$$= \iota_{E \to E'}{}^{-1}(X_{E'}(M)) = X_E(M). \qquad \square$$

We now have proved the following theorem.

**Theorem 9.3.2** (Avrunin–Scott). *Let $k$ be a field of characteristic $p$, let $G$ be a finite group, and let $M$ be a finitely generated $kG$-module. Then we have the following decomposition as a disjoint union:*

$$X_G(M) = \bigcup_{E \in \mathcal{I}} X_{G,E}(M)^+$$

*where $\mathcal{I}$ is a set of representatives of the distinct conjugacy classes of elementary abelian $p$-subgroups of $G$, and $X_{G,E}(M)^+ = \iota_E(X_E(M) \cap X_E^+)$.*

We may also derive the following important fact from the Subgroup Lemma.

**Corollary 9.3.3** (Subgroup Theorem). *If $H$ is any subgroup of the finite group $G$ and $M$ is a finitely generated $kG$-module, then*

$$\iota_{H \to G}^{-1}(X_G(M)) = X_H(M).$$

**Proof.** Use the Quillen Stratification and the basic lemma in the elementary abelian case.

**Exercise 9.3.1.** Complete the proof. □

## 9.4 The rank variety

Let $E$ be an elementary abelian $p$-group, $k$ a field of characteristic $p$, and $\Omega$ a field extension of $k$. As in Section 9.1 consider the set of points in $\Omega$

$$\Omega_E \cong E \otimes_{\mathbf{F}_p} \Omega \cong \mathrm{Hom}_k(E^*, \Omega) \cong \mathrm{Hom}_{k-\mathrm{alg}}(S(E^*), \Omega).$$

We can define a Zariski topology on the points in $\Omega$ by taking, as closed, sets of algebra homomorphisms vanishing on ideals, and in this case that topology coincides with the usual Zariski topology in $\Omega^{\mathrm{rank}\, E}$ which we may identify with $\Omega_E$. We may view $S(E^*)$ as an algebra of functions defined on $\Omega_E$ (but see the remarks below for how this relates to our identification of $S(E^*)$ with $H(E)/\mathrm{rad}\, H(E)$). Define the map $\Psi : \Omega_E \to X_E$ as in the proof of Theorem 9.1.7 by letting $\Psi(x)$ be the ideal of all functions vanishing at $x$. Assume in what follows that $\Omega$ is the algebraic closure of $k$. Then, as mentioned previously, the image $Y_E$ of $\Psi$ is the set of all closed points (maximal ideals) of $X_E$. Moreover, $Y_E$ is *dense* in $X_E$, and its intersection with any closed subset $V$ of $X_E$ is dense in $V$. (See Atiyah and MacDonald (1969, Chapter 5, Exercises 23–26).) $\Psi$ is certainly continuous, and the inverse image of any point is a finite set. Note that for $E$ cyclic of prime order, $Y_E = X_E$ (since every prime ideal is maximal).

Let $M$ be a $kE$-module. We wish to characterize the set $Y_E(M) = X_E(M) \cap Y_E$ which, as above, is dense in $X_E(M)$. Its inverse image $\Omega_E(M) = \Psi^{-1}(Y_E(M))$ is the locus in $\Omega_E$ of the homogeneous ideal $I_E(M)$ so it is a union of lines $\tilde{L}$ through the origin in $\Omega_E$. For each subgroup $F$ of $E$, we may view $\Omega_F$ as a subspace of $\Omega_E$ which is rational over $\mathbf{F}_p$; this is consistent via $\Psi$ with the one-to-one map $\iota_F : X_F \to X_E$, and $Y_E \cap \iota_F(X_F) = \iota_F(Y_F)$. We know from the Subgroup Lemma (Lemma 9.3.1) that
$$X_E(M) \cap \iota_F(X_F) = \iota_F(X_F(M)).$$
It follows that
$$Y_E(M) \cap \iota_F(Y_F) = \iota_F(Y_F(M))$$
and similarly
$$\Omega_E(M) \cap \Omega_F = \Omega_F(M).$$
Suppose in particular that $U$ is a cyclic subgroup of $E$. Then
$$X_U(M) = \begin{cases} \{0\} & \text{if } M \text{ is } kU\text{-free} \\ X_U & \text{otherwise.} \end{cases}$$
For $I_U(M)$ is a homogeneous radical ideal, and it is not hard to see that the only such ideal in $H(U)$ is either the ideal of all elements of positive degree or the zero ideal. The first case occurs if and only if $M$ is $kU$-projective, hence free. It follows that for $U$ a cyclic subgroup of $E$,
$$\iota_U(X_U) \subseteq X_E(M) \iff M \text{ is not } kU\text{-free}.$$
Given that $X_U = Y_U$, we may also say
$$\iota_U(Y_U) \subseteq Y_E(M) \iff M \text{ is not } kU\text{-free}$$
and
$$\Omega_U \subseteq \Omega_E(M) \iff M \text{ is not } kU\text{-free}.$$

Clearly, we also need to consider the lines through the origin $\tilde{L} \subseteq \Omega_E(M)$ (and their images $\Psi(\tilde{L}) \subseteq Y_E(M)$) which are not rational over $\mathbf{F}_p$. Each such line may be 'made rational' by changing the basis in $\Omega_E$ so that $\Omega_E = \Omega_{E'}$ for an appropriate $E' \cong E$. To do this, we need to introduce some terminology.

Let $R = \Omega E$, and let $J$ denote the augmentation ideal of $R$ (which is also its unique maximal ideal.) We saw in Section 2.3 that
$$\operatorname{Ext}^1_R(\Omega, \Omega) \cong \operatorname{Hom}_\Omega(J/J^2, \Omega) \cong \operatorname{Hom}_{\mathbf{F}_p}(E, \Omega) \cong H^1(E, \Omega)$$
and all this is consistent with the isomorphism $\Omega_E \cong J/J^2$ defined by $x \otimes a \mapsto a(x-1) \bmod J^2$. We shall show below in Appendix 1 that the

Bockstein homomorphism $\delta : \text{Ext}^1_R(\Omega,\Omega) \to \text{Ext}^2_R(\Omega,\Omega)$ does not actually depend on the identifications

$$\text{Ext}^1_R(\Omega,\Omega) \cong H^1(E,\Omega)$$
$$\text{Ext}^2_R(\Omega,\Omega) \cong H^2(E,\Omega).$$

Hence, we may view $\text{Ext}^*_R(\Omega,\Omega)$ modulo its radical as an algebra of functions on $J/J^2$, and this is completely consistent with what transpired above. (For $p$ odd, $\delta\,\text{Ext}^1_R(\Omega,\Omega) \cong (J/J^2)^*$ generates the symmetric algebra, but for $p = 2$ the issue of the Bockstein does not arise.)

Given a $d$-dimensional subspace $\tilde{L}$ of $J/J^2$, choose any $d$-dimensional subspace $L$ of $J$ which is complementary to $J^2$ and maps isomorphically on to $\tilde{L}$. Call the $\Omega$-subalgebra $S$ of $R$ generated by $L$ a *rank $d$ shifted algebra representing* $\tilde{L}$. If $\{x_1,\ldots,x_d\}$ is a basis for $L$, then it is not hard to see that $\{u_i = 1 + x_i \mid i = 1,\ldots,d\}$ is a basis for a rank $d$ elementary abelian $p$-subgroup $\tilde{F}$ of the multiplicative group $1 + J$, and $S \cong \Omega\tilde{F}$. We shall call $\tilde{F}$ a *shifted subgroup*. Completing to a basis modulo $J^2$ for $J/J^2$, we may form a complementary shifted subgroup $\tilde{F}'$ such that $E \cong \tilde{F} \times \tilde{F}'$, and $R \cong S \otimes_\Omega S'$ where $S' = \Omega\tilde{F}'$.

As above, consider the relationship between the ring $\text{Ext}^*_R(\Omega,\Omega)$ and the module $\text{Ext}^*_R(\Omega, M_\Omega)$ where $M_\Omega = M \otimes_k \Omega$. In principle, these structures depend on the identification $R = \Omega E$ and the diagonal map $R \to R \otimes R$ arising from the diagonal homomorphism $E \to E \times E$. However, as mentioned earlier in Section 8.1 the product may be described in terms of the Yoneda product which depends only on the ring $R$ and not on the diagonal map, i.e. not on the group $E$. (See also Appendix 2 below for an independent proof that the product does not depend on the diagonal map.) Similarly, we may consider the ring $\text{Ext}^*_S(\Omega,\Omega)$ and the module $\text{Ext}^*_S(\Omega, M_\Omega)$, and their relation to the above structures for $R$. In principle we could do this by trying to reproduce the entire theory of varieties as developed in the last two chapters, but fortunately the identifications $R \cong \Omega(F \times F') \cong S \otimes_\Omega S'$ allow us to fall back on the theory already developed.

Consider now the special case $d = 1$, i.e. $\tilde{L}$ is a line through the origin and $S = \Omega\tilde{U}$ where $\tilde{U}$ is a *cyclic* shifted subgroup. ($\tilde{U} = Gp<1+x>$ where $x \bmod J^2$ generates $\tilde{L}$.) We may now apply the previous considerations (over the field $\Omega$) to conclude that $\Omega_{\tilde{U}} = \tilde{L}$ and

$$\tilde{L} \subseteq \Omega_E(M_\Omega) \iff M_\Omega \text{ is not } S\text{-free}.$$

Since in these circumstances, $M$ is $k\tilde{U}$-projective if and only if $M_\Omega$ is $\Omega\tilde{U}$-projective, we have proved the following.

## The rank variety

**Theorem 9.4.1.** *Let $E$ be an elementary abelian p-group, $k$ a field of characteristic $p$, $\Omega$ its algebraic closure, and $M$ a $kE$-module. Then, with the above notation, $\Omega_E(M)$ consists of all one dimensional subspaces $\tilde{L}$ of $\Omega_E$ such that $M_\Omega$ is not $S$-free for each shifted subalgebra $S$ of $\Omega E$ representing $\tilde{L}$. $Y_E(M)$ consists of the images under $\Psi$ of all such $\tilde{L}$.*

**Remark.** Carlson originally defined the *rank variety* of a module $M$ over an elementary abelian $p$-group $E$ in terms of the freeness criterion stated above, and he conjectured that it was the same as the variety of the module. His conjecture was proved by Avrunin and Scott (1982), and their result is the essential content of the above theorem. Since Carlson, Avrunin, and Scott assume the base field $k$ is algebraically closed and work with the maximal ideal spectrum, their results are stated a bit differently.

The construction of shifted subgroups and the associated algebras does not depend on $\Omega$'s being algebraically closed. Such algebras allow us to derive a theorem of Carlson. (See Carlson (1983, Proposition 5.1) and Avramov (1987, Section 7) for generalizations.)

**Theorem 9.4.2.** *Let $E$ be an elementary abelian p-group of rank $r$, $k$ a field of characteristic $p$, and $M$ a finitely generated $kE$-module of complexity $d$. Then there is a finite extension $\Omega$ of $k$ and a decomposition $\Omega E \cong S \otimes_\Omega S'$ where $S$ is a shifted subalgebra of rank $d$, $S'$ is a shifted subalgebra of rank $r - d$, and $M_\Omega$ is $S'$-free. If $k$ is infinite, we may take $\Omega = k$.*

*The complexity $d$ may be characterized as the least integer with this property.*

**Proof.** Let $\mathfrak{a}$ be the annihilator of $\mathrm{Ext}^*_{kE}(M, M)$ in $H(E, k)$. The complexity $d$ of $M$ is $\dim |\mathrm{spec}\, H(E, k)/\mathfrak{a}|$. However, subject to the above identifications, $H(E, k)/\mathfrak{a}$ is generated as a $k$-algebra by the images modulo $\mathfrak{a}$ of a basis $\{\beta_1, \ldots, \beta_r\}$ for $(J/J^2)^*$. By the Noether Normalization Lemma, (Atiyah and MacDonald 1969, Chapter 5, Exercise 16), if $k$ is infinite, we may choose linear combinations $\xi_1, \ldots, \xi_d$ of the $\beta_j$ such that $H(E, k)/\mathfrak{a}$ is integral over the subalgebra generated by their images $\bar{\xi}_1, \ldots, \bar{\xi}_d$. If $k$ is finite, it may be necessary to make a finite extension $\Omega$ of $k$. However, in that case tensoring with $\Omega$ is essentially harmless with respect to what we want to prove.

Choose a shifted subalgebra $S'$ representing the $r - d$ dimensional linear subspace of $J/J^2$ defined by $\xi_1, \ldots, \xi_d$, and let $S$ be a complementary shifted subalgebra of rank $d$. As above, we may assume $S = kF, S' = kF'$, and $kE \cong k(F \times F')$. Then

$$\iota_{F'}(X_{F'}(M)) = \iota_{F'}(X_{F'}) \cap X_{F \times F'}(M)$$

and since $\iota_{F'}$ is one-to-one, in order to see that $M$ is a projective (hence free) $kF'$-module, it suffices to show that the intersection on the right is

{0}. However, since $\iota_{F'}X_{F'} = V(\xi_1, \ldots, \xi_d)$, it follows that the intersection is the variety defined by the ideal

$$\mathfrak{b} = (\xi_1, \ldots, \xi_d) + \mathfrak{a}$$

in $H(E, k)$. It suffices to show that the radical of that ideal is $\mathfrak{m} = (\beta_1, \ldots, \beta_r)$. However, it is not hard to see that $(\mathfrak{m}/\mathfrak{a})$ is the radical in $H(E, k)/\mathfrak{a}$ of the ideal $\mathfrak{b}/\mathfrak{a}$ since they both intersect the subring $k[\bar{\xi}_1, \ldots, \bar{\xi}_d]$ in the ideal of polynomials of positive degree. It follows that $\mathfrak{m} = \sqrt{\mathfrak{b}}$.

We leave it as an exercise for the reader to show that the complexity $d$ is characterized by this condition. □

**Remark.** All the 'varieties' we consider are homogeneous algebraic sets in affine $r$-space, so they determine algebraic sets in projective $r - 1$ space $\mathbf{P}^{r-1}$. Hence, to prove the theorem, we can simply appeal to the well known fact that, given an $i$-dimensional algebraic set in $\mathbf{P}^n$, there is a $j$-dimensional linear subspace which does not meet it if and only if $i + j < n$. In effect, the above argument reproves that assertion.

## Appendix 1: The Bockstein.

Let $G$ be a finite group, $k$ be a perfect field of characteristic $p > 0$, $R = kG$, and $J$ be the augmentation ideal. We shall show that the Bockstein $\delta : \mathrm{Ext}^1_R(k, k) \to \mathrm{Ext}^2_R(k, k)$ depends only on $R$ and $J$.

Define an $R$-free complex $\tilde{X}$ over $k$ as follows. Let $\tilde{X}_n = R \otimes J^{\otimes n}$. Let $\tilde{X}_0 = R \to k$ be the augmentation $\epsilon$, and for $n > 0$ define $\tilde{d}_n : \tilde{X}_n \to \tilde{X}_{n-1}$ by

$$\tilde{d}_n(1 \otimes v_1 \otimes v_2 \otimes \cdots \otimes v_n) = v_1 \otimes v_2 \otimes \cdots \otimes v_n$$
$$+ \sum_{i=1}^{n-1}(-1)^i 1 \otimes v_1 \cdots \otimes v_i v_{i+1} \otimes \cdots \otimes v_n$$

As in the case of the bar resolution (Section 2.3), it is easy to construct a contracting homotopy to show that $\tilde{X} \to k$ is a free resolution. In fact, this resolution is quite closely related to the bar resolution. Denote the latter by $X \to k$. Define $F_n : X_n \to \tilde{X}_n$ by

$$F_n([x_1|x_2|\ldots|x_n]) = 1 \otimes (x_1 - 1) \otimes (x_2 - 1) \otimes \cdots \otimes (x_n - 1).$$

We leave it to the reader to verify that this defines a map $F : X \to \tilde{X}$ of complexes over $k$. (It is a little more complicated than expected; there are some extra terms which cancel out.) Consider the 1-cochains

$$\mathrm{Hom}_R(R \otimes J, k) \cong \mathrm{Hom}(J, k).$$

A 1-cochain $\tilde{g} : J \to k$ is a 1-cocycle if and only if $u\tilde{g}(v) = \tilde{g}(uv)$ for $u, v \in J$. Since $uk = 0$ for $u \in J$, this amounts to saying $\tilde{g}(J^2) = 0$, i.e. $\tilde{g}$ lifts to a $k$-linear map $J/J^2 \to k$. As expected, composing with $F_2$ yields a homomorphism $g : G \to k$ where $g(x) = \tilde{g}(x - 1)$. Thus, we obtain an explicit description of the isomorphism $\mathrm{Hom}(J/J^2, k) \cong \mathrm{Hom}_{\mathbf{F}_p}(G/G_2, k)$.

Let $\alpha \in \mathrm{Ext}^1_R(k, k)$ be represented by $g$ or $\tilde{g}$ depending on the resolution. According to Exercises 3.4.1 and 3.4.2, $\delta\alpha$ is represented by the bar resolution 2-cocycle $h$ defined by

$$h(x|y) = -\left(\sum_{i=1}^{p-1} \frac{1}{p}\binom{p}{i} g(x)^i g(y)^{p-i}\right)^{1/p} \qquad x, y \in G.$$

But $h$ may be obtained by composing $\tilde{h}$ with $F_2$ where

$$\tilde{h}(1 \otimes u \otimes v) = -\left(\sum_{i=1}^{p-1} \frac{1}{p}\binom{p}{i} \tilde{g}(u)^i \tilde{g}(v)^{p-i}\right)^{1/p} \qquad u, v \in J,$$

so $\delta\alpha$ is also represented by $\tilde{h}$. Since $\tilde{h}$ is expressed purely in terms of the ring $R$ and the augmentation ideal $J$, we are done.

**Appendix 2: The Yoneda product.**

**Proposition 9.4.3.** *Let $G$ be $p$-group, $k$ a field of characteristic $p$, and $M$ a $kG$-module. The pairing*

$$\mathrm{Ext}^*_{kG}(k, k) \otimes \mathrm{Ext}^*_{kG}(k, M) \to \mathrm{Ext}^*_{kG}(k, M)$$

*depends only on the ring $kG$ and not on the diagonal homomorphism $kG \to kG \otimes kG$.*

**Proof.** Imbed $M$ in a cohomologically trivial module $Q$, and consider the short exact sequence of $kG$-modules

$$0 \to M \to Q \to M'' = Q/M \to 0.$$

(For example, let $Q = \mathrm{Hom}(kG, M)$, and identify $M = \mathrm{Hom}_{kG}(kG, M)$. As we mentioned earlier, there are two $kG$ module structures on $Q$ which produce isomorphic modules, and one of these, the module *induced* from the trivial module, has trivial cohomology in all positive degrees.) Because the cup product is consistent with the connecting homomorphism $\delta$ arising from this sequence, we have a commutative diagram

$$\begin{array}{ccc} H^q(G, k) \otimes H^r(G, M'') & \xrightarrow{(-1)^q \delta} & H^q(G, k) \otimes H^{r+1}(G, M) \\ \downarrow & & \downarrow \\ H^{q+r}(G, M) & \xrightarrow{\delta} & H^{q+r+1}(G, M) \end{array}$$

for each $q$ and $r \geq 0$. Since $H^r(G, M'') \to H^{r+1}(G, M)$ is an epimorphism (even an isomorphism for $r > 0$), we may argue inductively, so it suffices to show that the product depends only on $kG$ for $q = 0$. To do this consider the commutative diagram of groups

$$\begin{array}{ccc} G \times G & \xleftarrow{\Delta} & G \\ {\scriptstyle \mathrm{Id} \times \mathrm{triv}}\downarrow & & \downarrow{\scriptstyle \mathrm{Id}} \\ G \times 1 & \longleftarrow & G \end{array}$$

where the unlabelled arrow is defined by the diagram. In cohomology, this induces

$$\begin{array}{ccccc} H^q(G,k) \otimes H^0(G,M) & \xrightarrow{\times} & H^q(G \times G, k \otimes M) & \xrightarrow{\Delta^*} & H^q(G,M) \\ {\scriptstyle \cong}\uparrow & & \uparrow & & \uparrow \\ H^q(G,k) \otimes H^0(1, M^G) & \longrightarrow & H^q(G \times 1, k \otimes M^G) & \longrightarrow & H^q(G, M^G) \end{array}$$

It is not hard to see that the bottom arrows are isomorphisms which depend only on the ring $kG$ and not on the diagonal map $kG \to kG \otimes kG$. (Recall that $M^G$ can be defined as the set of all $x \in M$ killed by the augmentation ideal $J$ of $kG$ so it does not actually depend on $G$.) The vertical map on the right is that induced by the module inclusion $M^G \to M$ and it also depends only on the ring. Since the vertical map on the left is an isomorphism, the proposition follows. □

**Exercise 9.4.1.** Prove some of the unsubstantiated assertions in the above proof.

# 10
# Some related theorems

We round off our discussion of varieties and complexity with some consequences of the results in the previous two chapters and some other results employing ideas from commutative algebra.

## 10.1 The tensor product theorem and applications

**Theorem 10.1.1.** *Let $G$ be a group, $k$ be a field of characteristic $p > 0$, and $M$ and $N$ be finitely generated $kG$-modules. Then*
$$X_G(M \otimes N) = X_G(M) \cap X_G(N).$$

**Proof.** Consider the tower

$$X_G$$
$$\iota \downarrow$$
$$X_{G \times G}$$
$$\mu \downarrow$$
$$X_G \times X_G$$

where $\iota$ is induced by the diagonal homomorphism $\Delta : G \to G \times G$ and $\mu$ is induced by the two projections of $G \times G \to G$. ($\mu$ is not generally a homeomorphism or even one-to-one.) The composite $\mu \circ \iota$ is the diagonal homomorphism $X_G \to X_G \times X_G$, so the inverse image of $V \times W \subseteq X_G \times X_G$ is just $V \cap W$. Since $\Delta$ is a monomorphism, the Subgroup Theorem (Corollary 9.3.3) tells us that
$$\iota^{-1}(X_{G \times G}(M \otimes N)) = X_G(M \otimes N)$$
where $M \otimes N$ is considered a $k(G \times G)$-module in the usual way. Hence, it suffices to show that
$$X_{G \times G}(M \otimes N) = \mu^{-1}(X_G(M) \times X_G(N)).$$
To this end, let $\Omega$ be an algebraically closed extension of $k$ with transcendence degree at least $\dim H(G \times G)$, and consider the points in $\Omega$
$$\Omega_G = \mathrm{Hom}_{k\text{-alg}}(H^*(G, k), \Omega) \cong \mathrm{Hom}_{k\text{-alg}}(H(G), \Omega).$$

131

Consider the map $\pi : \Omega_G \to X_G$ defined by $\pi(\phi) = \text{Ker}\,\phi \cap H(G)$. By the assumption on $\Omega$, $\pi$ is onto (and it is also a continuous closed map). Let $\mathfrak{a}_G^*(M)$ be the set of all $\alpha \in H^*(G,k)$ such that $\alpha\,\text{Ext}_{kG}^*(M,M) = 0$ (or equivalently $\alpha\,\text{Id} = 0$). (If $p = 2$, $\mathfrak{a}_G^*(M) = \mathfrak{a}_G(M)$, but for $p > 2$, $\mathfrak{a}_G^*(M)$ may contain elements of odd degree.) Let

$$\Omega_G(M) = \{\phi\,|\,\phi(\mathfrak{a}_G^*(M)) = 0\} = \text{Hom}_{k\text{-alg}}(H^*(G,k)/\mathfrak{a}_G^*(M), \Omega)$$
$$= \text{Hom}_{k\text{-alg}}(H(G)/\mathfrak{a}_G(M), \Omega) = \text{Hom}_{k\text{-alg}}(H(G)/\sqrt{\mathfrak{a}_G(M)}, \Omega).$$

Clearly, $\Omega_G(M) = \pi^{-1}(X_G(M))$.

Let $G$ and $H$ be groups. The projections $G \times H \to G$ and $G \times H \to H$ induce a map $\tilde{\mu} : \Omega_{G \times H} \to \Omega_G \times \Omega_H$ which is a bijection (but not a homeomorphism). For, by the Künneth Theorem, we have

$$H^*(G \times H, k) \cong H^*(G,k) \otimes H^*(H,k)$$

and $\tilde{\mu}$ arises from the pair of algebra homomorphisms

$$H^*(G,k) \to H^*(G,k) \otimes H^*(H,k)$$
$$H^*(H,k) \to H^*(G,k) \otimes H^*(H,k).$$

That $\tilde{\mu}$ is a bijection follows from the fact that algebra homomorphisms $h : H^*(G) \otimes H^*(H) \to \Omega$ are exactly the maps defined by $h(\alpha \otimes \beta) = f(\alpha)g(\beta)$ for algebra homomorphisms $f : H^*(G) \to \Omega$ and $g : H^*(H) \to \Omega$. (The proof of this fact is a little tricky. Showing $h$ defined this way is an algebra homomorphism depends on the fact that elements of odd degree are sent to zero, since otherwise the sign introduced in the tensor product algebra would create difficulties.)

Consider the commutative diagram

$$\begin{array}{ccc} \Omega_{G \times H} & \longrightarrow & X_{G \times H} \\ \tilde{\mu} \downarrow & & \downarrow \mu \\ \Omega_G \times \Omega_H & \longrightarrow & X_G \times X_H \end{array}$$

where $\mu$ as above is induced from the projections $G \times H \to G$ and $G \times H \to H$. Let $M$ be a $kG$-module and $N$ a $kH$-module. It suffices to show that

$$X_{G \times H}(M \otimes N) = \mu^{-1}(X_G(M) \times X_H(N)).$$

Since the horizontal maps are onto, it suffices to prove the corresponding fact for $\tilde{\mu}$. However, $\tilde{\mu}^{-1}(\Omega_G(M) \times \Omega_H(N))$ is the subset of $\Omega_{G \times H}$ on which the ideal $\mathfrak{a}_G^*(M) \otimes H^*(H,k) + H^*(G,k) \otimes \mathfrak{a}_H^*(N)$ vanishes. Hence, the result follows from the following lemma.

**Lemma 10.1.2.** *Let $A$ and $B$ be $k$-algebras, and $M$ and $N$ modules over $A$ and $B$ respectively. Let $m \in M$ and $n \in N$, and let $I$ and $J$ be their respective annihilators. Then $A \otimes J + I \otimes B$ is the annihilator in $A \otimes B$ of $m \otimes n \in M \otimes N$.*

**Proof.** From the exact sequences

$$0 \to I \to A \to Am \to 0$$
$$0 \to J \to B \to Bn \to 0$$

we obtain the exact sequence

$$0 \to A \otimes J + I \otimes B \to A \otimes B \to Am \otimes Bn \to 0.$$

(See Cartan and Eilenberg (1956, Proposition II.4.3).) This completes the proof of the lemma and of the theorem.

Note that we could also have made the argument work by taking $\Omega$ to be the algebraic closure of $k$ and using density arguments. □

The rest of this section will be taken up with the construction of some interesting complexes. We start with the following theorem which is implicit in Carlson (1984).

**Theorem 10.1.3** (Carlson). *Let $G$ be a finite group, $k$ a field of characteristic $p > 0$, and $\mathfrak{a}$ a homogeneous ideal in $H(G)$. Then $V(\mathfrak{a}) = X_G(M)$ for an appropriate $kG$-module $M$.*

**Proof.** We have $\mathfrak{a} = (\alpha_1, \ldots, \alpha_r)$ for appropriate homogeneous elements $\alpha_i \in H(G)$, and $V(\mathfrak{a}) = V(\alpha_1) \cap \cdots \cap V(\alpha_r)$, so by the tensor product theorem, it suffices to prove the theorem for a 'hypersurface' $V(\alpha)$ where $\alpha$ is homogeneous of degree $n$. (For $p$ odd, $n$ will be even.) Let

$$\cdots \xrightarrow{d_{n+1}} X_n \xrightarrow{d_n} \cdots \longrightarrow X_0 \longrightarrow k \to 0$$

be a $kG$-projective resolution, and suppose $f : X_n \to k$ represents $\alpha$. Let $M$ be the *pushout* in the diagram

$$\begin{array}{ccc} X_n & \xrightarrow{d_n} & X_{n-1} \\ {\scriptstyle f}\downarrow & & \downarrow \\ k & \longrightarrow & M \end{array}$$

**Lemma 10.1.4.** *With the above notation,*

$$X_G(M) = V(\alpha).$$

**Proof.** Since

$$V(\alpha) = \bigcup_{E \text{ elem}} \iota_E(V(\mathrm{res}_E(\alpha))),$$

it follows from Theorem 8.3.1 that it suffices to prove the lemma for $G = E$, an elementary abelian $p$-group. However, by our analysis of the rank variety in Section 9.4, it suffices to show that $\Omega_E(M)$ is the subset $V'(\alpha)$ of $\Omega_E$ defined by $\alpha \in H(E)$ (where $\Omega$ is the algebraic closure of $k$). Hence, by Theorem 9.4.1 it suffices to show that $V'(\alpha)$ consists of those lines $\tilde{L}$ through the origin in $\Omega_E$ such that $M_\Omega$ is *not free* as an $S$-module for each cyclic shifted subalgebra $S$ representing $\tilde{L}$. In effect, this reduces the problem to proving the result for $E = U$ cyclic of order $p$ and $k$ algebraically closed. We now consider that case.

From our discussion of minimal resolutions in Section 2.4 we know that we have a decomposition of *complexes* $X = Y \oplus P$ where $P \to k$ is the special resolution for a cyclic group and $Y$ is a resolution of 0. (It follows that $Y \to 0$ has a contracting homotopy defined over $kU$, so at each stage $d_n Y_n$ is a direct summand of $Y_{n-1}$.) We leave it as an exercise for the reader to show in general that changing $f : X_n \to k$ by a coboundary $gd_n$, where $g : X_{n-1} \to k$, does not change (the isomorphism class of) the module $M$. It follows that we may assume $f$ is trivial on $Y_n$, and $\alpha = 0$ if and only if $f = 0$. For such an $f$, it is not hard to check that $M \cong M' \oplus Y_{n-1}/d_n Y_n$ where $M'$ is taken to be the pushout in the diagram

$$\begin{array}{ccc} kU & \xrightarrow{T} & kU \\ f' \downarrow & & \downarrow \\ k & \longrightarrow & M' \end{array}$$

Here, $f'$ is the restriction of $f$ to the summand $P_n = kU$, and for $p > 2$ and $n$ even, the boundary $P_n \to P_{n-1}$ is the trace map $kU \xrightarrow{T} kU$. (For $p = 2$, $T = 1 + u = u - 1$, so the parity of $n$ does not matter.) Since $Y_{n-1}/d_n Y_n$ is free, it suffices to show that $f' = 0$ if and only if $M'$ is not $kU$-free. If $f' = 0$, then it is easy to see that $k$ is a direct summand of $M'$ whence it is not free. If $f' \neq 0$, then define $i : k \to kU$ by $i(1) = T$ and $j : kU \to kU$ as $f'(1)$ Id. We can easily check that these maps present $kU$ as the required pushout $M'$.

This completes the proof of the lemma and of the theorem. □

**Exercise 10.1.1.**
   (a) With $X \to k$ a projective resolution as above, suppose

$$\begin{array}{ccccccccc} \cdots & \longrightarrow & X_n & \xrightarrow{d_n} & X_{n-1} & \longrightarrow & \operatorname{Coker} d_n & \longrightarrow & 0 \\ & & f \downarrow & & \downarrow & & \parallel & & \\ 0 & \longrightarrow & k & \longrightarrow & M & \longrightarrow & \operatorname{Coker} d_n & \longrightarrow & 0 \end{array}$$

commutes and has exact rows, and $f$ is a cocycle. Show that the relevant part of the diagram presents $M$ as a pushout.

(b) Suppose $\alpha_1$ and $\alpha_2$ are homogeneous elements in $H(G)$. Construct pushouts $M_1$ and $M_2$ for $\alpha_1$ and $\alpha_2$ as above using the resolution $X \to k$, and construct $M$ for $\alpha_1\alpha_2$ using the resolution $X \otimes X \to k$. Show that there is an epimorphism $M \to M_1 \oplus M_2$.

The following constructions are due to Benson and Carlson (1987).

It is not hard to check that the pushout diagram discussed above may be extended to a diagram

$$\begin{array}{ccccccccccc} \cdots & \to & X_n & \to & X_{n-1} & \to & X_{n-2} & \to & \cdots & \to & X_0 & \to & k & \to & 0 \\ & & \downarrow & & \downarrow & & \| & & & & \| & & \| & & \\ 0 & \to & k & \to & M & \to & X_{n-2} & \to & \cdots & \to & X_0 & \to & k & \to & 0 \end{array}$$

where the bottom row is also exact. (In fact, it represents the cohomology class $\alpha$ according to the Yoneda definition of $\operatorname{Ext}_{kG}^n(k,k)$.) Choose homogeneous elements $\alpha_i \in H(G)$ of degrees $n_i$ such that $H(G)$ is a finite (integral) extension of $k[\alpha_1,\ldots,\alpha_d]$. Then the radical of the ideal $(\alpha_1,\alpha_2,\ldots,\alpha_d)$ is the ideal $H(G)^+$ of elements of positive degree. For the latter ideal is maximal, and any element homogeneous $\xi$ in it satisfies a homogeneous polynomial equation of the form

$$\xi^r + f_1(\alpha_1,\ldots,\alpha_r)\xi^{r-1} + \cdots + f_r(\alpha_1,\ldots,\alpha_r) = 0$$

with coefficients in $(\alpha_1,\ldots,\alpha_r)$.

Each $\alpha_i$ defines a complex as above, and we let $Y(i)$ denote the complex

$$\cdots \to 0 \to M(i) \to X_{n_i-2} \to \cdots \to X_1 \to X_0 \to 0 \to \cdots$$

where we leave off the $k$'s on either end and extend by 0's in both directions. Note that

$$H_r(Y(i)) = \begin{cases} k & \text{for } r = 0, n_i - 1 \\ 0 & \text{otherwise.} \end{cases}$$

Form the complex

$$Y = Y(1) \otimes Y(2) \otimes \cdots \otimes Y(d).$$

By the Künneth Theorem, it follows that $H_r(Y)$ is either trivial or a direct sum of copies of $k$ with $H_r(Y) = k$ for $r = 0$ (the lowest non-trivial degree) and $r = \sum_i n_i - d$ (the highest nontrivial degree). In particular, $G$ acts trivially on $H_r(Y)$ for each $r$.

**Lemma 10.1.5.** *With the above notation, each constituent $Y_r$ of the complex $Y$ is projective.*

**Proof.** With the exception of the highest degree $r = \sum_i n_i - d$, each $Y_r$ is a sum of tensor products $Y_{r_1}(1) \otimes \cdots \otimes Y_{r_d}(d)$, at least one constituent $Y_{r_i}(i)$ of which is projective. It follows as in Section 8.1 that each such term is projective. For the highest degree term, we have

$$X_G(Y_r) = X_G(M(1) \otimes \cdots \otimes M(d))$$
$$= X_G(M(1)) \cap \cdots \cap X_G(M(r))$$
$$= V(\alpha_1) \cap \cdots \cap V(\alpha_r) = V(\alpha_1, \ldots, \alpha_r)$$

which, by the choice of the $\alpha_i$, is $\{0\}$. It follows that $Y_r$ is projective as claimed. □

Using the above complex, we may prove a special case of a theorem of Carlson.

**Theorem 10.1.6.** *Let $G$ be a finite group, $k$ a field of characteristic $p > 0$, and $M$ and $N$ finitely generated $kG$-modules. If $\operatorname{Ext}^n_{kG}(M,N) \neq 0$ for at least one $n > 0$, then $\operatorname{Ext}^n_{kG}(M,N) \neq 0$ for infinitely many $n > 0$.*

**Proof.** Let $Y$ be the complex constructed above and call its highest nontrivial degree $r_1$. Let $N \to Q$ be a $kG$-injective resolution, and consider the double complex $A = \operatorname{Hom}_{kG}(Y \otimes M, Q)$. As usual, there are two spectral sequences associated with $A$. If we take cohomology with respect to the first variable, use the fact that $\operatorname{Hom}_{kG}(-, Q)$ is exact, and then take cohomology with respect to the second variable, we obtain

$$E_2^{r,s} = \operatorname{Ext}^s_{kG}(H_r(Y) \otimes M, N).$$

If we reverse the order and use the fact that $Y$, and hence $Y \otimes M$, is $kG$-projective, we obtain instead

$$H^r(\operatorname{Ext}^s_{kG}(Y \otimes M, N)) = \begin{cases} 0 \text{ for } s > 0 \\ H^r(\operatorname{Hom}_{kG}(Y \otimes M, n)) \text{ for } s = 0 \end{cases}$$

for the $E_2$ term. Clearly, the second spectral sequence collapses, from which we may conclude that the total cohomology $H^n(A)$ vanishes for $n > r_1$.

Consider instead the first spectral sequence. Since $H_r(Y)$ is a direct sum of copies of $k$ with $G$ acting trivially, it follows that $H_r(Y) \otimes M$ is a direct sum of copies of $M$. Hence, $E_2^{r,s} = \operatorname{Ext}^s(H_r(Y) \otimes M, N)$ is just a direct sum of copies of $\operatorname{Ext}^s_{kG}(M,N)$. Also, since $H_{r_1}(Y) = k$, we have in the highest degree $E^{r_1,s} = \operatorname{Ext}^s_{kG}(M,N)$. If $\operatorname{Ext}^{s_1}_{kG}(M,N) \neq 0$ but $\operatorname{Ext}^s_{kG}(M,N) = 0$ for $s > s_1$, there would be no way to kill off the corner

$E_2^{r_1,s_1}$ in the spectral sequence, so it would follow that $H^n(A) \ne 0$ for $n = r_1 + s_1 > r_1$. □

An analogue of the the complex used in the above proof may also be constructed over **Z**. (See Benson and Carlson (1987, 1991).) The argument then works just as well provided $M$ is **Z**-free. However, by dimension shifting, it is then possible to derive the result for any finitely generated modules.

**Exercise 10.1.2.** Generalize the above argument to show that there cannot be gaps in $\mathrm{Ext}_{kG}^s(M, N)$ of size larger than $r_1$ unless it vanishes for all $s > 0$.

Benson and Carlson (1987) discuss another variation of their construction which has some useful consequences. To describe it, fix $\alpha \in H(G)$, and construct as above an exact sequence

$$0 \to k \to M \to X_{n-2} \to \cdots \to X_0 \to k \to 0$$

where $X_G(M) = V(\alpha)$ and each $X_i$ is projective. We may compose this sequence with itself repeatedly to obtain a complex

$$\cdots \to X_0 \to M \to X_{n-2} \to \cdots \to X_0 \to M \to \cdots \to X_0 \to k \to 0.$$

This provides an acyclic complex over $k$, which we denote $W \to k$, which is *periodic* of period $n = \deg \alpha$. All the components $W_i$ are projective except for $i = kn - 1$ in which case we have $M$. Suppose now that $N$ is a finitely generated $kG$-module of complexity $d$. Then as above we may choose $\alpha_1, \alpha_2, \ldots, \alpha_d$ such that $H(G)/\mathfrak{a}_G(N)$ is integral over the subalgebra generated by the images of the $\alpha_i$. It follows that the radical of $(\alpha_1, \ldots, \alpha_d) + \mathfrak{a}_G(N)$ is the ideal of elements of positive degree. Thus,

$$V(\alpha_1, \ldots, \alpha_d) \cap X_G(N) = \{0\}.$$

Choose as previously a complex over $k$, $W(i) \to k$, for each $\alpha_i, i = 1, \ldots d$. Then

$$U = W(1) \otimes W(2) \otimes \cdots \otimes W(d) \otimes N \to k \otimes \cdots \otimes k \otimes N = N$$

is an acyclic complex over $N$, and since

$$X_G(M(1) \otimes \cdots \otimes M(d) \otimes N) = X_G(M(1)) \cap \cdots \cap X_G(M(d)) \cap X_G(N)$$
$$= V(\alpha_1, \ldots, \alpha_d) \cap X_G(N) = \{0\},$$

it follows as above that $U \to N$ is a projective resolution of $N$. $U$ is '$d$-tuply periodic' in the sense that it is a tensor product with $N$ of $d$ complexes, each of which is periodic.

It follows from the above construction that a non-projective module $N$ has complexity 1 if and only if it has a periodic projective resolution, as proved in Proposition 8.4.4. It is interesting to note that Alperin invented the concept of complexity in order to generalize the periodic case. He was able to construct examples of multiply periodic resolutions, but he was unable at that stage to make a definition based on that idea, so he resorted to using the growth rate of $\dim \operatorname{Ext}^i_{kG}$. After the introduction of varieties and further development of the theory, Benson and Carlson were able to confirm his original intuition.

## 10.2 Varieties and corestriction

Experience shows that the corestriction homomorphism from a subgroup to a group carries relatively little information; often it is even trivial. (See Lemma 6.3.4.) However, Carlson (1987) proved a remarkable theorem relating the sum total of *all corestrictions* from subgroups of index divisible by $p$ to the kernel of the *restriction* to the centre of a $p$-Sylow subgroup.

As usual, let $G$ be a finite group, and $k$ a field of characteristic $p$. Then, for any abelian subgroup $A$ of $G$, $H(A)/\operatorname{rad} H(A)$ is a polynomial ring, so

$$\mathfrak{p}_A = \operatorname{rad}(\operatorname{Ker} \operatorname{res}_{G \to A}) = \{\xi \in H(G) \mid \operatorname{res}_{G \to A} \xi \text{ is nilpotent}\}$$

is a prime ideal of $H(G)$. Where we need to indicate the dependence of this ideal on the group $G$, we shall use the notation $\mathfrak{p}_{G,A} = \mathfrak{p}_A$. Note also that for any subgroup $H$ of $G$, Proposition 4.2.4 implies that $\operatorname{Im} \operatorname{cor}_{H \to G}$ is an ideal in $H(G)$.

**Theorem 10.2.1** (Carlson). *Let $G$ be a finite group, and let $Z$ be the centre of $P$, a $p$-Sylow subgroup of $G$. Let*

$$J_G = \sum_{p \mid (G:H)} \operatorname{Im} \operatorname{cor}_{H \to G}.$$

*Then*

$$\iota_Z(X_Z) = V(J_G) = \text{the closed subspace of } X_G \text{ defined by } J_G;$$

*i.e.* $\mathfrak{p}_Z = \operatorname{rad} J_G$.

The proof is adapted from Evens and Feshbach (1989) which generalizes Carlson's theorem to compact Lie groups.

**Proof.** We first reduce to the case that $G$ is a $p$-group. By Proposition 4.2.5, $\operatorname{res}_{G \to P}$ is a monomorphism on to a direct summand of $H(P)$.

**Lemma 10.2.2.** $\mathrm{res}_{G \to P}(J_G) = \mathrm{res}_{G \to P}(H(G)) \cap J_P$.

**Proof.** Choose $H$ so that $p \mid (G:H)$. Then by the double coset formula (Theorem 4.2.6), $\mathrm{res}_{G \to P} \mathrm{cor}_{G \to H} \alpha$ is a sum of terms of the form

$$\mathrm{cor}_{gHg^{-1} \cap P \to P} \mathrm{res}_{P \to gHg^{-1} \cap P} g^* \alpha$$

for appropriate double coset representatives $g \in G$. To see that these terms lie in $J_P$, it suffices to show that $p \mid (P : gHg^{-1} \cap P)$. However, since $p \mid (G : gHg^{-1})$ and $p \nmid (G : P)$, the diagram

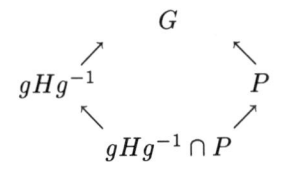

$$\begin{array}{c} G \\ \nearrow \quad \nwarrow \\ gHg^{-1} \qquad\qquad P \\ \nwarrow \quad \nearrow \\ gHg^{-1} \cap P \end{array}$$

shows that $p \mid (P : gHg^{-1} \cap P)$. Thus, $\mathrm{res}(J_G) \subseteq J_P$.

Conversely, let $\xi \in J_P \cap \mathrm{res}(H(G))$. Then, $\xi$ is stable (Corollary 4.2.7) and

$$\mathrm{res}_{G \to P} \mathrm{cor}_{P \to G} \xi = (G : P)\xi.$$

Since $\xi$ is a sum of elements of the form $\mathrm{cor}_{H \to P} \eta$ where $p \mid (P : H)$, then so is $(G : P)^{-1} \xi$. Hence, by the transitivity of transfer,

$$\xi = \mathrm{res}_{G \to P} \mathrm{cor}_{P \to G} (G/P)^{-1} \xi \in \mathrm{res}(J_G). \qquad \square$$

$\square$

**Lemma 10.2.3.** $\mathrm{res}_{G \to P}(\mathfrak{p}_{G,Z}) = \mathrm{res}_{G \to P}(H(G)) \cap \mathfrak{p}_{P,Z}$.

**Proof.** This follows since res is transitive and $\mathrm{res}_{G \to P}$ is injective. $\square$

We now assume that $G$ is a finite $p$-group.

In the statement of the theorem, the centre $Z$ of $P$ may be replaced by its maximal elementary abelian $p$-subgroup $Z'$. For modulo radicals $\mathrm{res}_{Z \to Z'}$ is a monomorphism so, $\mathfrak{p}_Z = \mathfrak{p}_{Z'}$ and $\iota_Z(X_Z) = \iota_{Z'}(X_{Z'})$. Henceforth, we shall just write $Z$ for $Z'$.

**Lemma 10.2.4.** $J_G \subseteq \mathfrak{p}_Z$, so $V(J_G) \supseteq \iota_Z(X_Z)$.

**Proof.** First, assume $H \not\supseteq Z$. Then, since $Z$ is central, no conjugate of $H$ contains $Z$. Thus, the double coset formula implies that the composition $\mathrm{res}_{G \to Z} \mathrm{cor}_{H \to G} = 0$ since, by Lemma 6.3.4, $\mathrm{cor}_{K \to Z} = 0$ for all proper subgroups $K \subset Z$.

Suppose alternatively $H \supseteq Z$. In this case, the double coset formula yields $\mathrm{res}_{G \to Z} \mathrm{cor}_{H \to G} = (G : H) \mathrm{res}_{H \to Z}$, which is zero if $p \mid (G : H)$. It follows that $J_G \subseteq \mathfrak{p}_Z$. $\square$

The rest of the proof of the theorem is concerned with showing

$$V(J_G) \subseteq \iota_Z(X_Z).$$

To this end, let $\mathfrak{p} \in X_G$, and let $\mathfrak{q} \in X_E$ be a source for $\mathfrak{p}$ ($E$ a vertex) as in Section 9.3. Assume moreover that $\mathfrak{p} \notin \iota_Z(X_Z)$ so (by the definitions of *source* and *vertex*) $E$ is not a subgroup of $Z$, i.e. $E$ is not central. Let $H = C_G(E)$, which by the previous remark is a proper subgroup of $G$. We shall show $\mathfrak{p} \notin V(\operatorname{Im} \operatorname{cor}_{H \to G}) \supseteq V(J_G)$.

As in Section 9.1, modulo its radical, $H(E)$ is a symmetric algebra on the $k$-subspace spanned by $E\hat{\ }$. Consider, as before, the product $\epsilon_E \in H(E)$ of the *non-zero* elements of $E\hat{\ }$. For each $\xi \in H(E)$, let $\tilde{\xi} = N_{E \to H}(\xi)$. Then

$$\operatorname{res}_{H \to E} \tilde{\xi} = \xi^{(H:E)} \tag{10.1}$$

but

$$\operatorname{res}_{H \to E'} \tilde{\epsilon}_E = 0 \tag{10.2}$$

for any elementary abelian $p$-subgroup $E'$ of $H$ which does not contain $E$. These facts follow from Theorem 6.1.1; the first follows from (N4), and the second follows from the double coset formula (N3) since any conjugate of $E'$ intersects $E$ in a proper subgroup and $\operatorname{res} \epsilon_E = 0$ for such a subgroup.

Let $W = N_G(E)/C_G(E)$. Then, by Corollary 9.1.8, $\epsilon_E^q = Tr_W(\mu)$ for some $\mu \in H(E)$ and some power $q$. Since $\epsilon_E$ is invariant under $W$, we have

$$Tr_W(\operatorname{res}_{H \to E}(\tilde{\epsilon}_E \tilde{\mu})) = Tr_W(\epsilon_E^{(H:E)} \mu^{(H:E)})$$
$$= \epsilon_E^{(H:E)} Tr_W(\mu)^{(H:E)} = \epsilon_E^r \tag{10.3}$$

where $r = (q+1)(H:E)$.

Let $\tilde{\alpha} = \tilde{\epsilon}_E \tilde{\mu} \in H(H)$. We shall show that $\operatorname{cor}_{H \to G} \tilde{\alpha} \notin \mathfrak{p}$, so that $\mathfrak{p} \notin V(\operatorname{Im} \operatorname{cor}_{H \to G})$. For this, it suffices in turn to show that

$$\operatorname{res}_{G \to E} \operatorname{cor}_{H \to G} \tilde{\alpha} \notin \mathfrak{q}.$$

However, by Theorem 4.2.6, we have the double coset decomposition

$$\operatorname{res}_{G \to E} \operatorname{cor}_{H \to G} \tilde{\alpha} = \sum \operatorname{cor}_{gHg^{-1} \cap E \to E} g^*(\operatorname{res}_{H \to H \cap g^{-1}Eg} \tilde{\alpha}).$$

Since $\tilde{\epsilon}_E$ is a factor of $\tilde{\alpha}$, (10.2) yields $\operatorname{res}_{H \to H \cap g^{-1}Eg} \tilde{\alpha} = 0$ if $H \cap g^{-1}Eg \not\supseteq E$, i.e. if $g^{-1}Eg \neq E$. Hence, the only double cosets in the sum are those coming from left (right) cosets of $H = C_G(E)$ in $N_G(E)$. Thus, the sum is just $Tr_W(\operatorname{res}_{H \to E} \tilde{\alpha})$, and applying (10.3) gives

$$\operatorname{res}_{G \to E} \operatorname{cor}_{H \to G} \tilde{\alpha} = Tr_W(\operatorname{res}_{H \to E} \tilde{\epsilon}_E \tilde{\mu}) = \epsilon_E^r.$$

However, since $\mathfrak{q}$ is a vertex for $\mathfrak{p}$ with source $E$, it follows that $\mathfrak{q} \in X_E^+$ whence no power of $\epsilon_E$ vanishes at $\mathfrak{q}$. Hence, $\operatorname{res}_{G \to E}(\operatorname{cor}_{H \to G} \tilde{\alpha}) \notin \mathfrak{q}$ as claimed.

This completes the proof of the theorem. □

**Exercise 10.2.1.** Let $G$ be finite, and let $Z$ be the centre of a $p$-Sylow subgroup $P$ of $G$. Let $J_G'$ be the sum of the ideals $\operatorname{Im}\operatorname{cor}_{H \to G}$ where $H = C_P(x)$ for $x \in P$, $x \notin Z$, $x$ of order $p$. Prove

$$\iota_Z(X_Z) = V(J_G) = V(J_G').$$

Hint: In the above argument, with $G = P$, $H = C_G(E) \subseteq C_G(x) \neq G$ for some non-central element $x \in E$.

## 10.3 Depth

As usual, let $G$ be a finite group and $k$ a field of characteristic $p > 0$. If $M$ and $N$ are finitely generated $kG$-modules, we shall consider the $H(G)$-module $\operatorname{Ext}_{kG}^*(M, N)$. The variety $X_G(M)$ tells us something about the structure of this module, but there are other invariants of modules which play an important role in commutative algebra. We shall consider here one such invariant, namely the *depth*.

Recall a few definitions. Let $R$ be a graded (strictly) commutative ring with $R^0 = k$ and $R^i = 0$ for $i < 0$. Note that $R$ is necessarily a $k$-algebra and that the ideal $\mathfrak{m}$ of elements of positive degree is maximal. For convenience, call such a ring a *G-algebra over $k$*. Assume in addition that $R$ is noetherian, and let $A$ be a finitely generated graded $R$-module. An $R$-sequence on $A$ (relative to $\mathfrak{m}$) is a sequence of homogeneous elements $\mu_1, \mu_2, \ldots, \mu_n \in \mathfrak{m}$ such that for each $i = 0, 1, \ldots, n-1$, $\mu_i$ does not annihilate any non-trivial element of $A/(\mu_1 A + \cdots + \mu_{i-1} A)$.

The $R$-depth of $A$ (relative to $\mathfrak{m}$) is defined as the *common* length of all maximal $R$-sequences (from $\mathfrak{m}$) on $A$. See Kaplansky (1974, Section 3-1)—which uses the synonym 'grade' for 'depth'—for a discussion of the basic theory. Some theorems about depth are stated for local rings using the unique maximal ideal or for other rings using the Jacobson radical, but these theorems also hold for noetherian $G$-rings relative to the unique graded maximal ideal. Concentration on homogeneous elements introduces no essential difficulties. Finally, the quotient modules $A/(\mu_1 A + \cdots + \mu_{i-1} A)$ are never trivial for graded modules over $G$-rings. See Stanley (1978), which is the source for much of our terminology, for discussion of the graded case.

The following inequality is an immediate consequence of the definition.

$$R\text{-depth of } (A_1 \oplus A_2) \leq \min(R\text{-depth of } A_1, R\text{-depth of } A_2). \quad (10.4)$$

In fact, equality holds in (10.4). This follows, for example, from the following characterization of depth (relative to $\mathfrak{m}$). The $R$-depth of $A$ is the least positive integer $n$ such that $\mathrm{Ext}_R^n(k,A) \neq 0$. (See Kaplansky (1974, Appendix 3-1).)

Duflot (1981) derived an inequality for depth for the *equivariant cohomology* of a finite group acting on a space for the trivial module $k$. Here we give another proof of Duflot's theorem for group cohomology (where the space is a point) but for general modules.

**Theorem 10.3.1** (Duflot). *Let $k$ be a field of characteristic $p > 0$. Let $G$ be a finite group and $M$ and $N$ finitely generated $kG$-modules. Let $E$ be an elementary abelian p-subgroup of $G$ contained in the centre of a p-Sylow subgroup which acts trivially on $\mathrm{Hom}_k(M,N)$. If $\mathrm{Ext}_{kG}^*(M,N) \neq 0$, then*

$$H(G)\text{-depth of } \mathrm{Ext}_{kG}^*(M,N) \geq \text{ rank of } E.$$

**Proof.** We first reduce to the case $G$ is a $p$-group.

Let $P$ be a $p$-Sylow subgroup of $G$. By Proposition 4.2.5 we have a direct sum decomposition

$$H^*(P, \mathrm{Hom}_k(M,N)) \cong \mathrm{Ext}_{kP}^*(M,N) \cong \mathrm{Ext}_{kG}^*(M,N) \oplus T \qquad (10.5)$$

where everything is an $H(G)$-module. By (10.4), it follows that

$$H(G)\text{-depth of } \mathrm{Ext}_{kG}^*(M,N) \geq H(G)\text{-depth of } \mathrm{Ext}_{kP}^*(M,N).$$

To complete the reduction to $P$, it suffices to show that

$$H(G)\text{-depth of } \mathrm{Ext}_{kP}^*(M,N) = H(P)\text{-depth of } \mathrm{Ext}_{kP}^*(M,N).$$

This follows from the following lemma.

**Lemma 10.3.2.** *Let $R$ and $S$ be noetherian $G$-algebras over $k$ with $S$ an integral extension of $R$. Let $\mathfrak{m}$ be the maximal ideal of elements of positive degree in $R$ and $\mathfrak{M}$ the corresponding maximal ideal in $S$. Finally, let $A$ be a finitely generated graded $S$-module. Then*

$$R\text{-depth of } A \text{ relative to } \mathfrak{m} = S\text{-depth of } A \text{ relative to } \mathfrak{M}.$$

**Proof.** It is clear that any $R$-sequence (from $\mathfrak{m}$) on $A$ is an $S$-sequence (from $\mathfrak{M}$) on $A$. Let $\mu_1, \mu_2, \ldots, \mu_n \in \mathfrak{m}$ be a maximal $R$-sequence on $A$. It suffices to show that it is also a maximal $S$-sequence. By maximality, any $\mu \in \mathfrak{m}$ annihilates some non-trivial element of $\overline{A} = A/(\mu_1 A + \cdots + \mu_n A)$. By Kaplansky (1974, Theorem 82) there is a non-trivial element $\overline{\alpha} \in \overline{A}$ such

that $\mathfrak{m}\bar{a} = 0$. Let $\xi \in \mathfrak{M}$ be homogeneous. Then $\xi$ satisfies an integral relation
$$\xi^k + \nu_1 \xi^{k-1} + \cdots + \nu_k = 0$$
with $\nu_i \in R$, and by picking out homogeneous components we may assume each $\nu_i \in \mathfrak{M} \cap R = \mathfrak{m}$. It follows that $\xi^k \bar{a} = 0$ for at least one $k > 0$. If we choose $k$ minimal, it follows that $\xi$ annihilates the non-trivial element $\xi^{k-1}\bar{a} \in \bar{A}$. Thus, the given sequence cannot be extended by any element of $\mathfrak{M}$, and is a maximal $S$-sequence as required. $\square$

Suppose now that $G$ is a $p$-group and $E$ is an elementary abelian subgroup of rank $d$ contained in its centre, which acts trivially on $\mathrm{Hom}_k(M,N)$. Let $\{x_1, \ldots, x_d\}$ be an $\mathbf{F}_p$-basis for $E$, and let $\{\xi_1, \ldots, \xi_d\}$ be a dual basis for $E^*$ which we may view as embedded in $H(E)$ as a subspace of degree 2. Finally, let $\zeta_i = N_{E \to G}(\xi_i)$ for $i = 1, \ldots, d$. We shall prove that $\zeta_1, \zeta_2, \ldots, \zeta_d$ is an $H(G)$-sequence on $A = \mathrm{Ext}^*_{kG}(M,N)$ provided the latter is non-trivial.

Let $\bar{A}_j = A/(\zeta_1 A + \cdots + \zeta_{j-1} A)$ (with the convention $\bar{A}_0 = A$). Note that $\zeta_j$ does not annihilate a non-trivial element of $\bar{A}_j$ if and only if no positive power of $\zeta_j$ does so. Fix an $i$ and let $C$ be the cyclic central subgroup of $G$ generated by $x = x_i$. By Theorem 6.1.1
$$\mathrm{res}_{G \to C}(N_{E \to G}(\xi_i)) = \xi^q$$
where $\xi = \mathrm{res}_{E \to C}(\xi_i)$ spans $H^2(C, k)$ and $q = (G : E)$. The group extension
$$1 \to C \to G \to G/C \to 1$$
induces two spectral sequences
$$H^*(G/C, H^*(C,k)) = H^*(G/C, k) \otimes H^*(G, k) \Rightarrow H^*(G,k) \qquad (10.6)$$
and
$$H^*(G/C, \mathrm{Ext}^*_{kC}(M,N)) \Rightarrow \mathrm{Ext}^*_{kG}(M,N) \qquad (10.7)$$
with the latter a module over the former. (See the proof of Lemma 8.3.2.) $\xi^q$ (or any power of it) is a universal cycle in $E_2^{0,*}$ for the spectral sequence (10.6). As in the proof of Lemma 7.4.4, it is not hard to see that both spectral sequences stop, i.e. there is an $r$ such that $E_r = E_\infty$. Let $B = \sum_{n=0}^{\infty} F^{n-r+1} \mathrm{Ext}^n_{kG}(M,N)$. The elements of $B$ represent all elements of $\bar{B} = \sum_{t < r} E_\infty^{*,t}$ in (10.7), i.e. all terms *below* the line $t = r$. Since we can use multiples of $r$ just as well as $r$, there is no loss of generality in assuming that $r$ is a multiple of $2q$ so that
$$\xi^{r/2} = \mathrm{res}_{G \to C} \zeta_i^{r/2q}.$$

The argument below is not sensitive to replacing $\zeta_i = \zeta$ by some power, so we shall suppose $r = 2q$.

Multiplication by $\xi^{r/2}$ is an isomorphism of the $t$th row of the $E_2$-term of the spectral sequence (10.7) on to its $(t+r)$th row. This is clear for $t > 0$ by the known multiplicative structure for the cohomology of $C$, and, under the hypothesis of trivial $C$-action, it is also true for $t = 0$ since

$$\operatorname{Ext}^0_{kC}(M, N) \cong \operatorname{Hom}(M, N)^C \cong \operatorname{Hom}_k(M, N).$$

Since $\xi^{r/2}$ is a universal cycle in the spectral sequence (10.6), it follows that multiplication by that element is an isomorphism of one row on to another for each term $E_{r'}$ including $r' = \infty$. It follows that $E_\infty = k[\xi^{r/2}] \otimes \tilde{B}$ and since $\zeta \in H(G)$ represents $\xi^{r/2}$, it follows that $\operatorname{Ext}^*_{kG}(M, N) = k[\zeta] \otimes B$.

Consider now the elements $\zeta_j$ for $j \neq i$. Since $C \subseteq E$ and since $\xi_j$ is the inflation of some element of $\xi'_j \in H^2(E/C, k)$, it follows from Theorem 6.1.1 that

$$\zeta_j = N_{E \to G}(\xi_j) = N_{E \to G}(\inf \eta'_j) = \inf N_{E/C \to G/C}(\eta'_j).$$

Being inflations, $\zeta_j, j \neq i$, represent elements in the bottom row ($t = 0$) of the $E_\infty$ term of the spectral sequence (10.6). The fact that $\zeta_1, \ldots, \zeta_d$ is an $H(G)$-sequence on $\operatorname{Ext}^*_{kG}(M, N)$ now follows from the following argument.

Let $R = \operatorname{Im} \inf_{G/C \to G} \cap H(G)$, and let $\mathfrak{q} = (\mu_1, \ldots, \mu_j)$ be a homogeneous ideal $R$ consisting of elements of positive degree. (For example, let $\mathfrak{q}$ be generated by a subset of $\{\zeta_j \mid j \neq i\}$.) As above,

$$A = \operatorname{Ext}^*_{kG}(M, N) \cong R[\zeta] \otimes_R B = B[\zeta].$$

**Lemma 10.3.3.** *With the above notation, $\zeta$ does not annihilate any nontrivial element of $A/\mathfrak{q}A$.*

**Proof.** Let $\alpha \in A$, and suppose $\zeta \alpha = \sum_i \mu_i \alpha_i$ where $\alpha_i \in A$. Each $\alpha_i$ may be written uniquely:

$$\alpha_i = \alpha'_i \zeta + \gamma_i$$

where $\alpha'_i \in A$ and $\gamma_i \in B$. It follows that

$$\zeta(\epsilon - \sum_i \mu_i \alpha'_i) = \sum_i \mu_i \gamma_i \in B,$$

which implies that both sides are 0. Since $\zeta$ does not annihilate any element of $A = B[\zeta]$, it follows that

$$\epsilon = \sum_i \mu_i \alpha'_i \equiv 0 \bmod \mathfrak{q}A$$

as claimed.

This completes the proof of the theorem. □

The following useful fact about $R$-sequences is referred to in Stanley (1978, p. 63).

**Proposition 10.3.4.** *Let $R$ be a $G$-algebra over $k$, and let $A$ be a finitely generated graded $R$-module. Any $R$-sequence $\theta_1, \theta_2, \ldots, \theta_n$ on $A$ forms an algebraically independent set, and $A$ is a free module over $k[\theta_1, \theta_2, \ldots, \theta_n]$.*

In general, $A$ is not of finite rank over $k[\theta_1, \ldots, \theta_n]$. In the case when it is, $A$ is called a *Cohen–Macaulay module*.

**Proof.** We argue as in Cartan and Eilenberg (1956, Proposition 5.2), but by induction on $n$.

Let $n = 1$. Choose homogeneous elements $\overline{\alpha}_i$, $i \in \mathcal{I}$, whose images in $A/\theta_1 A$ form a $k$-basis. Then $\{\alpha_i \mid i \in \mathcal{I}\}$ is a $k[\theta_1]$-basis for $A$. For to check that it spans, let $\alpha = \sum_i c_i \alpha_i + \theta_1 \alpha'$ and note that $\alpha'$ has lower degree than $\alpha$. (By finite generation, there is a lowest possible degree.) Similarly, to check that it is linearly independent, use the fact that $\theta_1$ is not a zero divisor in $A$, and argue inductively on the minimal degree of a coefficient in a linear relation.

Suppose $n > 1$. By induction, $A$ is free over $k[\theta_1, \ldots, \theta_{n-1}]$, and $\overline{A} = A/(\theta_1 A + \cdots + \theta_{n-1} A)$ is free over $k[\theta_n]$. Choose $\{\alpha_i \mid i \in \mathcal{I}\}$ such that its image in $\overline{A}$ is a $k[\theta_n]$-basis. Let $S = \{\theta_n{}^j \alpha_i \mid j \in \mathbf{N}, i \in \mathcal{I}\}$, and let $F$ be the free $k[\theta_1, \ldots, \theta_{n-1}]$-module on this set. The canonical morphism $f : F \to A$ yields an exact sequence

$$0 \to K = \operatorname{Ker} f \to F \to A \to C = \operatorname{Coker} f \to 0.$$

Let $I = (\theta_1, \ldots, \theta_{n-1})$. Tensoring over $k[\theta_1, \ldots, \theta_{n-1}]$ with $k$ yields

$$F/IF \to A/IA \to C/IC \to 0,$$

but by construction $F/IF \to A/IA$ is an isomorphism. Hence, $C/IC = 0$, which, as noted above, implies $C = 0$. Thus, $F \to A$ is an epimorphism, and we obtain

$$\operatorname{Tor}_1^{k[\theta_1, \ldots, \theta_{n-1}]}(A, k) \to K/IK \to F/IF \to A/IA \to 0.$$

Since $A$ is free over $k[\theta_1, \ldots, \theta_{n-1}]$ by the induction hypothesis, the Tor term vanishes. However, since $F/IF \to A/IA$ is an isomorphism, this means that $K/IK = 0$, i.e. $K = 0$, and $F \to A$ is an isomorphism. □

**Corollary 10.3.5.** *Let $G$ be a finite group and $E$ an elementary abelian subgroup contained in the centre of a $p$-Sylow subgroup of $G$. Let $k$ be a field of characteristic $p > 0$, and let $M$ and $N$ be finitely generated $kG$-modules such that $E$ acts trivially on $\operatorname{Hom}_k(M, N)$. Then we can find elements $\zeta_1, \zeta_2, \ldots, \zeta_d \in H^*(G, k)$ of even degree such that $H^*(E, k)$ is a finite module over the polynomial subring $k[\operatorname{res} \zeta_1, \ldots, \operatorname{res} \zeta_d]$ and $\operatorname{Ext}_{kG}^*(M, N)$ is a free module over the polynomial ring $k[\zeta_1, \ldots, \zeta_d]$.*

**Exercise 10.3.1.** Suppose the $p$-Sylow subgroups of $G$ are abelian. Show that $H^*(G, k)$ is Cohen–Macaulay module over $H(G)$.

# References

Alperin, J. L. (1977). Periodicity in groups. *Ill. J. Math.*, **21**, 776–783.

Alperin, J. L. and Evens, L. (1981). Representations, resolutions, and Quillen's dimension theorem. *J. Pure Appl. Alg.*, **22**, 1–9.

Alperin, J. L. and Evens, L. (1982). Varieties and elementary abelian subgroups. *J. Pure Appl. Alg.*, **26**, 221–227.

Atiyah, M. F. and MacDonald, I. G. (1969). *Introduction to commutative algebra*. Addison-Wesley, Reading, MA.

Atiyah, M. F. and Wall, C. T. C. (1967). Cohomology of groups. In *Algebraic number theory* (eds. J. W. S. Cassels and A. Fröhlich), pp. 94–115. Academic Press, London.

Avramov, L. L. (1987). Modules of finite virtual projective dimension. Report, Department of Mathematics, University of Stockholm, Sweden.

Avrunin, G. S. (1981). Annihilators of cohomology modules. *J. Alg.*, **69**, 150–154.

Avrunin, G. S. and Scott, L. L. (1982). Quillen stratification for modules. *Invent. Math.*, **66**, 277–286.

Baer, R. (1934). Erweiterung von Gruppen und ihre Automorphismen. *Math. Zeit.*, **38**, 375–416.

Barr, M. and Rinehart, G. S. (1966). Cohomology as the derived functor of derivations. *Trans. Amer. Math. Soc.*, **16**, 416–426.

Benson, D. J. (1984). *Modular representation theory: new trends and methods*. Springer-Verlag, Berlin.

Benson, D. J. (1991). *Representations and cohomology I: basic representation theory of finite groups and associative algebras*. Cambridge Studies in Advanced Mathematics **30**. Cambridge University Press, Cambridge.

Benson, D. J. (in press). *Representations and cohomology II: cohomology of groups and modules*. Cambridge Studies in Advanced Mathematics. Cambridge University Press, Cambridge.

Benson, D. J. and Carlson, J. F. (1987). Complexity and multiple complexes. *Math. Zeit.*, **195**, 221–238.

Benson, D. J. and Carlson, J. F. (1991). Projective resolutions and Poincaré duality complexes. To appear.

Benson, D. J. and Evens, L. (1990). Group homomorphisms inducing isomorphisms in cohomology. *Commun. Alg.*, **18**, 3447–3451.

Beyl, F. R. (1981). The spectral sequence of a group extension. *Bull. Sci. Math. Sér. 2*, **105**, 407–434.

Bieri, R. (1976). *Homological dimension of discrete groups.* Queen Mary College Mathematics Notes. Mathematics Department, Queen Mary College, London.

Bourbaki, N. (1972). *Elements of mathematics, commutative algebra* (English translation). Addison Wesley, Reading, MA.

Brown, K. S. (1982). *Cohomology of groups.* Springer-Verlag, New York.

Carlson, J. F. (1983). The varieties and the cohomology ring of a module. *J. Alg.*, **85**, 104–143.

Carlson, J. F. (1984). The variety of an indecomposable module is connected. *Invent. Math.*, **77**, 291–299.

Carlson, J. F. (1987). Varieties and transfer. *J. Pure Appl. Alg.*, **44**, 99–105.

Cartan, H. and Eilenberg, S. (1956). *Homological algebra.* Princeton University Press, Princeton, NJ.

Cassels, J. W. S. and Fröhlich, A. (eds.) (1967). *Algebraic number theory.* Academic Press, London.

Chapman, G. R. (1982). The cohomology ring of a finite abelian group. *Proc. Lond. Math. Soc.*, **45**, 564–576.

Charlap, L. S. and Vasquez, A. T. (1966). The cohomology of group extensions. *Trans. Amer. Math. Soc.*, **124**, 24–40.

Charlap, L. S. and Vasquez, A. T. (1969). Characteristic classes for modules over groups. *Trans. Amer. Math. Soc.*, **137**, 533–549.

Chase, S. U., Harrison, D. K., and Rosenberg, A. (1964). Galois theory and Galois cohomology of commutative rings. *Mem. Amer. Math. Soc.*, **12**, 15–33.

Chouinard, L. (1976). Projectivity and relative projectivity over group rings. *J. Pure Appl. Alg.*, **7**, 278–302.

Curtis, C. W. and Reiner, I. (1981). *Methods of representation theory,* vol. I. Wiley, New York.

Diethelm, T. (1985). The mod $p$ cohomology rings of the nonabelian split metacyclic $p$-groups. *Arch. Math.*, **44**, 29–38.

Duflot, J. (1981). Depth and equivariant cohomology. *Comment. Math. Helv.*, **56**, 627–637.

Eckmann, B. (1953). Cohomology of groups and transfer. *Ann. Math.*, **58**, 481–493.

Eilenberg, S. and Mac Lane, S. (1947a). Cohomology theory in abstract groups, I. *Ann. Math.*, **48**, 51–78.

Eilenberg, S. and Mac Lane, S. (1947b). Cohomology theory in abstract groups, II; group extensions with a non-abelian kernel. *Ann. Math.*, **48**, 326–341.

Eisenbud, D. (1980). Homological algebra on a complete intersection with an application to group representations. *Trans. Amer. Math. Soc.*, **260**, 35–64.

Evens, L. (1961). The cohomology ring of a finite group. *Trans. Amer. Math. Soc.*, **101**, 224–239.

Evens, L. (1963). A generalization of the transfer map in the cohomology of groups. *Trans. Amer. Math. Soc.*, **108**, 54–65.

Evens, L. (1965). On the Chern classes of representations of finite groups. *Trans. Amer. Math. Soc.*, **115**, 180–193.

Evens, L. (1975). The spectral sequence of a finite group extension stops. *Trans. Amer. Math. Soc.*, **212**, 269–277.

Evens, L. and Feshbach, M. (1989). Carlson's theorem on varieties and transfer. *J. Pure Appl. Alg.*, **57**, 39–45.

Evens, L. and Friedlander, E. M. (1982). On $K_*(\mathbf{Z}/p^2\mathbf{Z})$ and related homology groups. *Trans. Amer. Math. Soc.*, **270**, 1-46.

Evens, L. and Jackowski, S. (1990). A note on the subgrooup theorem in cohomological complexity theory. *J. Pure Appl. Alg.*, **65**, 25–28.

Golod, E. S. and Safarevic, I. R. (1964). On class field towers (in Russian). *Izv. Akad. Nauk. SSSR*, **28**, 261–272. English translation in *AMS Transl.* (2), **48**, 91–102.

Gruenberg, K. W. (1970). *Cohomological topics in group theory*. Springer-Verlag, Berlin.

Hall, M. (1959). *The theory of groups*. Macmillan, New York.

Hilton, P. J. and Stammbach, U. (1971). *A course in homological algebra*. Springer-Verlag, New York.

Hochschild, G. and Serre, J. P. (1953). Cohomology of group extensions. *Trans. Amer. Math. Soc.*, **74**, 110–134.

Huebschmann, J. (1989). The mod $p$ cohomology rings of metacyclic groups. *J. Pure Appl. Alg.*, **60**, 53–103.

Huppert, B. (1967). *Endliche Gruppen*, I. Springer-Verlag, Berlin.

Jackowski, S. (1978). Group homomorphisms inducing isomorphisms in cohomology. *Topology*, **17**, 303–307.

Kaloujnine, L. and Krasner, M. (1948). Le produit complet des groups de permutations et le problème d'extension des groupes. *C. R. Acad. Sci. Paris*, **227**, 806–808.

Kaloujnine, L. and Krasner, M. (1950). Produit complet des groups de permutations et le problème d'extensions des groupes I. *Acta Sci. Math. Szeged*, **13**, 208–230.

Kaloujnine, L. and Krasner, M. (1951a). Produit complet des groups de permutations et le problème d'extensions des groupes II. *Acta Sci. Math. Szeged*, **14**, 39–66.

Kaloujnine, L. and Krasner, M. (1951b). Produit complet des groups de permutations et le problème d'extensions des groupes III. *Acta Sci. Math. Szeged*, **14**, 69–82.

Kaplansky, I. (1974). *Commutative rings*. University of Chicago Press, Chicago, IL.

Karpilovsky, G. (1987). *The Schur multiplier*. Clarendon Press, Oxford.

Kroll, O. (1985). A representation theoretical proof of a theorem of Serre. Aarhus University Preprint Series, no. 33.

Lewis, G. (1968). The integral cohomology rings of groups of order $p^3$. *Trans. Amer. Math. Soc.*, **132**, 501–529.

Lyndon, R. (1948). The cohomology theory of group extensions. *Duke Math. J.*, **15**, 271–292.

Mac Lane, S. (1949). Cohomology theory in abstract groups, III. *Ann. Math.*, **50**, 736–761.

Mac Lane, S. (1963). *Homology*. Springer-Verlag, Berlin.

Mac Lane, S. (1978). Origins of the cohomology of groups. *Enseign. Math.*, **24**, 1–29.

Okuyama, T. and Sasake, H. (1990). Evens' norm map and Serre's theorem on the cohomology algebra of a $p$-group. *Arch. Math.*, **54**, 331–339.

Polya, G. and Read, R. C. (1987). *Combinatorial enumeration of groups, graphs, and chemical compounds*. Springer-Verlag, Berlin.

Priddy, S. B. and Wilkerson C. (1985). Hilbert's Theorem 90 and the Segal Conjecture for elementary abelian $p$-groups. *Amer. J. Math.*, **107**, 775–785.

Quillen, D. (1971a). A cohomological criterion for $p$-nilpotence. *J. Pure Appl. Alg.*, **1**, 361–372.

Quillen, D. (1971b). The spectrum of an equivariant cohomology ring, I. *Ann. Math.*, **94**, 549–572.

Quillen, D. (1971c). The spectrum of an equivariant cohomology ring, II. *Ann. Math.*, **94**, 573–602.

Quillen, D. and Venkov, B. B. (1972). Cohomology of finite groups and elementary abelian subgroups. *Topology*, **11**, 317–318.

Ratcliffe, J. G. (1980). Crossed homomorphsims. *Trans. Amer. Math. Soc.*, **257**, 73–89.

Rinehart. G. S. (1969). Satellites and cohomology. *J. Alg.*, **12**, 295–329.

Roquette, P. (1967). On class field towers. In *Algebraic number theory* (eds. J. W. S. Cassels and A. Fröhlich), pp. 231–249. Academic Press, London.

Schreier, O. (1926a). Über Erweiterungen von Gruppen, I. *Monatsh. Math. Phys.*, **34**, 165–180.

Schreier, O. (1926b). Über Erweiterungen von Gruppen, II. *Abh. Math. Sem. Hamburg Univ.*, **4**, 321–346.

Schur, I. (1904). Über die Darstellung der endlichen Gruppen durch gebrochene lineare Substitutionen. *J. reine angew. Math.*, **127**, 20–50.

# References

Schur, I. (1907). Untersuchungen über die Darstellung der endlichen Gruppen durch gebrochene lineare Substitutionen. *J. reine angew. Math.*, **132**, 85–137.

Schur, I. (1911). Über die Darstellung der symmetrischen und der alternierenden Gruppe durch gebrochene lineare Substitutionen. *J. reine angew. Math.*, **139**, 155–250.

Serre, J. P. (1965a). Sur la dimension cohomologique des groupes profinis. *Topology*, **3**, 413–420.

Serre, J.-P. (1965b). *Cohomologie Galoisienne.* Lecture Notes in Mathematics **5**. Springer-Verlag, Berlin.

Serre, J.-P. (1979). *Local fields.* Graduate Texts in Mathematics **67**, translated by M. J. Greenberg. Springer-Verlag, Berlin.

Serre, J. P. (1987). Une relation dans la cohomologie des $p$-groupes. *C. R. Acad. Sc. Paris*, Sér. I, **304**, 587–590.

Stallings, J. (1965). Homology and central series of groups. *J. Alg.*, **2**, 170–181.

Stammbach, U. (1973). *Homology in group theory.* Lecture Notes in Mathematics **359**. Springer-Verlag, Berlin.

Stanley, R. P. (1978). Hilbert functions of graded algebras. *Adv. Math.*, **28**, 57–83.

Steenrod, N. (1962). *Cohomology operations* (revised by D. B. A. Epstein), Annals of Mathematics Studies **50**. Princeton University Press, Princeton, NJ.

Swan, R. G. (1960a). Induced representations and projective modules. *Ann. Math.*, **71**, 552–578.

Swan, R. G. (1960b). The $p$-period of a finite group. *Ill. J. Math.*, **4**, 341–346.

Swan, R. G. (1962). Projective modules over group rings and maximal orders. *Ann. Math.*, **76**, 55–61.

Swan, R. G. (1969). Groups of cohomological dimension one. *J. Alg.*, **12**, 585–610.

Tate, J. (1963). Nilpotent quotient groups. *Topology*, **3**, suppl. 1, 109–111.

Thomas, C. B. (1986). *Characteristic classes and the cohomology of finite groups.* Cambridge Studies in Advanced Mathematics **9**. Cambridge University Press, Cambridge.

Townsley Kulich, L. G. (1988). Investigations of the integral cohomology ring of a finite group. Ph. D thesis. Northwestern University, Evanton, IL.

Venkov, B. B. (1959). Cohomology algebras for some classifying spaces (in Russian). *Dokl. Akad. Nauk SSSR*, **127**, 943–944.

Wall, C. T. C. (1961). Resolutions for extensions of groups. *Proc. Camb. Phil. Soc.*, **57**, 251–255.

Weir, A. (1955). Sylow $p$-subgroups of the classical groups over finite fields with characteristic prime to $p$. *Proc. Amer. Math. Soc.*, **6**, 529-533.

# Table of notation

$(A : B; d)$ 119
$(E \otimes_{\mathbf{F}_p} \Omega)^+$ 115
$1 \int \alpha$: wreath product class 51
$A \uparrow_H^G$ 36
$B^{[q]}$ 118
$E\widehat{\phantom{x}}$ 109
$E_r^{p,q}$: spectral sequence term 71
$F$-isomorphism 121
$H(G)$ 93
$H^*(G, \text{Hom}(N, M))$ 94
$H^+(G, M)$ 87
$H^2(G, \mathbf{Z})$ 29
$H^n(G, M)$ 2
$H^{ev}(G, k)$ 87
$H_n(G, M)$ 2
$I_G(M)$ 94
$M \otimes_{kG} N$ 2
$M^G$: invariants of $G$ on $M$ 1
$M^{\otimes(G/H)}$: tensor induced module 46, 48
$M_G$: coinvariants of $G$ on $M$ 2
$N_{H \to G}(\alpha)$: norm 57
$S \int H$: wreath product 47
$S_k(V)$: symmetric algebra over $k$ 117
$V(\mathfrak{a})$: closed subspace defined by an ideal 94
$X_E^+$ 111
$X_G$: variety of a group 93

$X_G(M)$: variety of a module 94
$X_{G,E}(M)^+$ 124
$X_{G,E}^+$ 111
$[x_1|x_2|\ldots|x_n]$ 7
$\text{Ext}_{kG}^*(M, N)$ 2
$\text{Ext}_{kG}$ for $k$ a field 94
$\mathfrak{a}_G(M)$ 94
$\mathfrak{p}_E$ 109
$\text{Hom}_{kH}(kG, A)$: coinduced module 36
$\Omega_E$: points in $\Omega$ 114
$\Omega_E(M)$ 125
$\Phi(X, Y)$ 31
$\text{Tor}_*^{kG}(M, N)$ 2
$\text{Tr}_W$: trace 116
$\alpha \times \beta$ 22
$\beta$: Bockstein 28
$\text{cor}_{H \to G}$: corestriction 39
$\delta$: Bockstein 28
$\epsilon_F$ 110
$\iota_{H \to G}$ 97
$\text{res}_{G \to H}$: restriction 35
$\text{Der}(G, M)$: derivations 7
$\text{InDer}(G, M)$: inner derivations 9
$\times$: cross product 22, 32
$\zeta_F$ 110
$cx_G(M)$: complexity of $M$ 103
$f \times g$ 17
$kG \otimes_{kH} A$: induced module 36

# Index

$1\int\alpha$: wreath product class 51

$(A:B;d)$ 119
abelian group, cohomology of 18
abelian group, cohomology ring of 32
abelian group, elementary 32
acyclic complex 5
Alexander–Whitney map 25
Alperin vi, 13, 86, 100, 104, 138
$\alpha \times \beta$ 22
amalgamated subgroup 37
$A\uparrow_H^G$ 36
artinian ring 11
associated graded ring 72
Atiyah v, 86
augmentation 2
augmentation ideal 2
Avrunin 100
Avrunin–Scott 122, 124, 127

Baer v
Baer sum 10
bar resolution 7
bar resolution, normalized 8
Barr 11
base ring, independence of 3
Benson viii, 13, 122, 135
$\beta$: Bockstein 28
Bockstein homomorphism 28, 30, 126, 128
$B^{[q]}$ 118
Brown vii

Carlson 13, 95, 127, 133, 135, 136, 138

Cartan–Eilenberg sign convention 4
Cayley–Hamilton 123
Chapman 33
Charlap-Vasquez 83
Chase 116
Chern classes viii
Chern classes of a unitary representation 59
Chouinard 104
class field theory v, 77
class field tower 14
classifying space vi, 92
closed map 114
closed points 115
cocycles, normalized 8
Cohen–Macaulay module 145
cohomology ring of a direct product 32
coinduced module 36, 74
coinvariants 1
collapse of a spectral sequence 73
commutative graded ring 21, 24, 87
complexity 103, 127, 138
connected varieties 133
connecting homomorphism 26, 43
contracting homotopy 5, 8
$\text{cor}_{H\to G}$: corestriction 39
corestriction 39, 138
cross product 22
cross product algebra v
cross product, associativity of 23
cross product, commutativity of 23
crossed homomorphism 6
cup product 21, 22
cup product, associativity of 24
cup product, commutativity of 24

$cx_G(M)$: complexity of $M$ 103
cyclic group 5
cyclic group, cohomology ring of 26, 29
cyclic group, products for 25
cyclic group, resolution for 5

Dedekind domain 1, 49
$\delta$: Bockstein 28
depth 141
derivation 28, 38
derivation, inner 9
derivation of a group 6
derivation of an algebra 21
derivations 10
diagonal action 1
diagonal chain map 25
diagonal homomorphism 1, 22
Diethelm 86
differential graded algebra 21
dimension 103, 105
direct limit 121
direct product, cohomology of 17
direct product, cohomology ring of 32
double complex 3, 69
double coset 41
double coset formula for corestriction 41
double coset formula for tensor induction 48
Duflot 142

Eckmann v, 36
Eckmann–Shapiro lemma 36
edge homomorphism 76
Eilenberg v, 10
Eilenberg–Mac Lane space 82
Eisenbud 106
elementary abelian group, cohomology ring of 32
elementary abelian $p$-group 63
elementary abelian $p$-group, rank of 103

elementary abelian $p$-groups 32, 62
elementary abelian $p$-subgroup 100, 109
elementary abelian $p$-subgroups v, 118
$(E \otimes_{\mathbf{F}_p} \Omega)^+$ 115
$\epsilon_F$ 110
equivalence of group extensions 9
equivalence of splittings 11
$E_r^{p,q}$: spectral sequence term 71
$E\hat{\phantom{E}}$ 109
$\mathrm{Ext}_{kG}$ for $k$ a field 94
$\mathrm{Ext}^*_{kG}(M,N)$ 2
exterior algebra 32

factor set 9
Feshbach 138
filtered complex 69
filtered ring 91
first spectral sequence 73
$F$-isomorphism 121
Frattini series 78
Frattini subgroup 78
free group 6
free product, cohomology of 11
free product with amalgamation 37
Freudenthal v
Frobenius endomorphism 31
$f \times g$ 17
fundamental exact sequence 77

Galois ring extension 116
generic point 109
$\mathfrak{a}_G(M)$ 94
$\mathfrak{p}_E$ 109
Golod 14
group extension v, 9, 72
group extension, spectral sequence of 72
growth rate 105
Gruenberg vii, 7
Grün's 2nd Theorem 42

$H^2(G, \mathbf{Z})$ 29

# Index

Hall subgroup 40
Harrison 116
$H^{ev}(G, k)$ 87
$H(G)$ 93
$H^*(G, \text{Hom}(N, M))$ 94
$H^+(G, M)$ 87
Hilbert Basis Theorem 87
Hilbert *Nullstellensatz* 116
$H^n(G, M)$ 2
$H_n(G, M)$ 2
Hochschild–Serre 82
$\text{Hom}_{kH}(kG, A)$: coinduced module 36
Hopf v
Huebschmann 86
Hurevic v

$I_G(M)$ 94
indecomposable projective 11
induced module 36, 45
inductive limit 121
inflation 4, 76
injective hull 13
injective modules 12
inner derivation 9
invariants 1
inverse limit 119
$\iota_{H \to G}$ 97

Jackowski 86, 122
Jacobson radical 93

Kaloujnine–Krasner 45
$kG \otimes_{kH} A$: induced module 36
Krull dimension v, 103, 105
Künneth Theorem 17, 22, 69, 132

Lewis 67
LHS spectral sequence 69, 89
locally closed 112

Mac Lane v, 4, 10
Mayer–Vietoris sequence 38
$M_G$: coinvariants of $G$ on $M$ 2

$M^G$: invariants of $G$ on $M$ 1, 98
minimal primes 110
minimal resolution 13, 14, 50, 107, 134
monomial module 45
monomial representation 47, 57
$M^{\otimes(G/H)}$: tensor induced module 46, 48
$M \otimes_{kG} N$ 2
multiply periodic resolution 137

Nakaoka 50, 69
$N_{H \to G}(\alpha)$: norm 57
nil radical 93
Noether Normalization Lemma 105
noetherian module 87
non-triviality of cohomology groups 59
non-triviality of restriction 58
norm viii, 57
normal basis theorem 116
normalized bar resolution 8
normalized cocycles 8

$\Omega_E$: points in $\Omega$ 114
$\Omega_E(M)$ 125

pairing 21
perfect field 31, 128
periodic module 107
periodic resolution 137
$p$-group 14
$p$-group, minimal resolution for 14
$p$-group, projective module 14
$\Phi(X, Y)$ 31
$p$-nilpotent groups 79
Poincaré series 15, 20, 105
points in $\Omega$ 115, 124, 131
Polya 45
presentations 7
Priddy 116
primary component 40
primary decomposition 111
prime ideal spectrum 93

principal ideal domain 16
projective cover of a module 12
projective limit 119
projective module 3, 103
projective representations v
pushout 135

quasi-Frobenius ring 12
Quillen v, 101, 103, 109, 112, 115, 117, 118
Quillen homeomorphism 118

radical, nil 93
rank variety 127, 134
reduced word 7
$\text{res}_{G \to H}$: restriction 35
resolution 5
resolution, bar 7
resolution for a cyclic group 5
resolution for a direct product 17
resolution for a semi-direct product 18
resolution from a presentation 7
resolution, multiply periodic 137
restriction 4, 35, 76
restriction, non-triviality of 58
restriction of a module to a subgroup, notation for 98
Rinehart 11
Roquette 14
Rosenberg 116

Safarevic 14
Schreier v
Schur v, 77
Schur multiplier v
Schur–Zassenhaus Theorem 43, 77
second spectral sequence 73
semi-direct product 18
semi-direct product, cohomology of 50
Serre 100
Serre's Theorem 64
Shapiro's Lemma 36

shifted algebra 126
shifted subalgebra 134
shifted subgroup 122, 126
sign convention 4, 49, 54, 72, 81
$S_k(V)$: symmetric algebra over $k$ 117
socle 13
source 114
spectral sequence 69
spectral sequence, collapse 73
spectral sequence of a group extension 69, 72
spectral sequence, product in 80
spectrum vii, 93
split group extension 11, 43
splitting 11
stable element 42
Stallings 79
Stammbach vii
Stanley 141
Steenrod reduced powers viii, 51
Steenrod's Lemma 52
stratification for groups 112
stratification for modules 122
Subgroup Lemma 122
Subgroup Theorem 124
Swan v, 42
$S \int H$: wreath product 47
Sylow subgroup 40
symmetric algebra 63, 109
symmetric group, cohomology of viii, 43

Tate v, 79
Tate cohomology vii
Tate–Nakayama Theorem 77
tensor induced module 45
tensor product 2
tensor product complex 17
tensor product theorem for varieties 131
$\text{Der}(G, M)$: derivations 7
$\text{InDer}(G, M)$: inner derivations 9
Thompson v

## Index

×: cross product 22, 32
$\operatorname{Tor}_*^{kG}(M,N)$ 2
Townsley Kulich 33
$\operatorname{Tr}_W$: trace 116
trace 116
transfer 39
transgression 77
trivial action 2

Universal Coefficient Theorem 29

variance 3
variety 93
variety of a module vi, 94
Venkov 92, 101
vertex 114
$V(\mathfrak{a})$: closed subspace defined by an ideal 94

Weir 45
Wilkerson 116
Witt vectors 31
wreath product 20, 45, 46, 47
wreath product, cohomology of 50

$[x_1|x_2|\ldots|x_n]$ 7
$X_E^+$ 111
$X_G$: variety of a group 93
$X_{G,E}^+$ 111
$X_{G,E}(M)^+$ 124
$X_G(M)$: variety of a module 94

Yoneda product 96, 126, 129

Zariski topology 93, 124
$\zeta_F$ 110

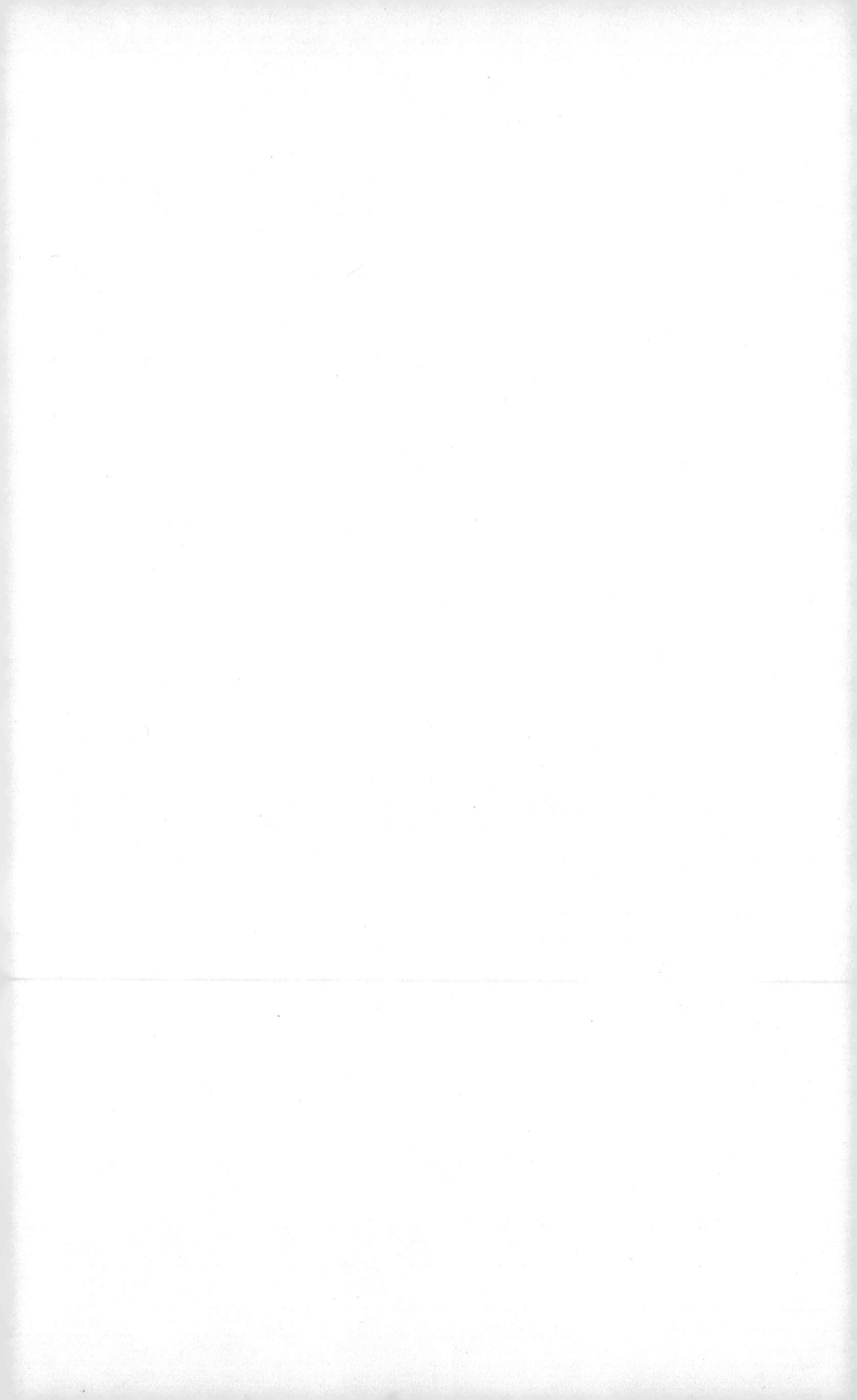